ARCHIPELAGO

ARCHIPELAGO:

THE ISLANDS OF INDONESIA

From the Nineteenth-Century Discoveries of Alfred Russel Wallace to the Fate of Forests and Reefs in the Twenty-First Century

Gavan Daws and Marty Fujita

Prologue by Edward O. Wilson

Epilogue by John C. Sawhill

Published in Association with The Nature Conservancy

UNIVERSITY OF CALIFORNIA PRESS BERKELEY LOS ANGELES LONDON

University of California Press
Berkeley and Los Angeles, California

University of California Press, Ltd.
London, England

Library of Congress Cataloging-in-Publication Data

Daws, Gavan.
Archipelago : the islands of Indonesia : from the nineteenth-century discoveries of Alfred Russel Wallace to the fate of forests and reefs in the twenty-first century / Gavan Daws and Marty Fujita ; prologue by Edward O. Wilson ; epilogue by John C. Sawhill.
p. cm.
"Published in association with the Nature Conservancy."
Includes bibliographical references and index.
ISBN 0-520-21576-1 (alk. paper)
1. Wallace, Alfred Russel, 1823–1913—Journeys—Malay Archipelago. 2. Naturalists—England—Biography. 3. Zoogeography—Malay Archipelago. 4. Natural history—Malay Archipelago. 5. Evolution (Biology)—Malay Archipelago. 6. Natural selecton. I. Fujita, Marty, 1954– II. Title.
QH31.W2 D28 1999
508.598—ddc21 99-22769
CIP

Printed in Hong Kong

08 07 06 05 04 03 02 01 00 99
10 9 8 7 6 5 4 3 2 1

The paper used in this publication meets the minimum requirements of ANSI/NISO Z39.48-1992 (R 1997) (Permanence of Paper).⊗

CONTENTS

MAPS

ACKNOWLEDGMENTS

This book is in great part the result of the generous contribution of the time and expertise, and hearts and minds, of many. First and foremost, we gratefully acknowledge ARCO, and especially President and Resident Manager of ARCO Indonesia, Leon Codron. With Leon's belief in our book's goal to raise awareness around the world of the need to conserve Indonesia's biological heritage, ARCO generously provided financial support to The Nature Conservancy to underwrite all expenses for the preparation of the manuscript. This financial support has enabled The Nature Conservancy to channel all profits generated from the sale of this book to conservation efforts in Indonesia.

Countless hours of library research on subjects ranging from the biology of birds of paradise to the history of Victorian museums and cabinets were devoted by Anne Keary. Anne managed to provide us with all the information we requested while we collaborated on the text, often from locations that spanned the Pacific Ocean, and we gratefully acknowledge her heroic efforts. We are also indebted to Jane Camerini, who provided us with rich details and insights on Alfred Russel Wallace. Such insights are not found in dusty references, but only through the benefit of years of diligent research into the life and times of Wallace.

Many of The Nature Conservancy's staff in Indonesia and its Asia-Pacific Regional Office provided vision, facts, and moral and administrative support for our efforts. We are particularly grateful to Carol Fox, Suzanne Case, Kelvin Taketa, Amy Bruno, Wayne Klockner, Rili Djohani, Darwina Wijayanti, Duncan Neville, and Donald Bason. Nathan Lau and Steve Miyashiro provided expert computer and internet assistance, which kept us connected over thousands of miles of Pacific Ocean. We are most grateful for the commitment and many kindnesses bestowed upon us by members of The Nature Conservancy Indonesia Program's Advisory Board, especially A. R. Ramly, Joe Bartlett, Beni Wahju, Shanti Poesposoetjipto, and James Castle. We hope this book will help them further the cause of conservation in Indonesia. For their noble and steadfast efforts to conserve the reefs and rain forests of Indonesia, and the support they have given to The Nature Conservancy's Indonesia Program, we also thank former Indonesian Minister of Forestry Ir. Djamaloedin Soerjohadikoesoemo and former Indonesian Minister of Environment Ir. Sarwono Kusumaatmadja.

We sought the expertise of many ecologists, science historians, and environmentalists. Through e-mail interviews, phone calls, faxes and letters, they generously responded from around the world with clarity, thoughtfulness, a wealth of personal experiences and research, intriguing field stories, and true collegiality. Their voices, featured throughout the book, brought life and individual perspectives to many of the topics we present. A resounding thanks to Suraya Afiff, Michael Alvard, Peter Ashton, Bruce Beehler, Nora Bynum, Jane Camerini, Gordon Corbet, Rili Djohani, the Earl of Cranbrooke, John Hill, Jonathan Hodge, Bob Johannes, Kuswata Kartawinata, Cheryl Knott, Tim Laman, Ernst Mayr, John Mitani, James Moore, Duncan Neville, David Quammen, A. J. S. Reid, Steven Siebert, Eli Silver, Dan Simberloff, Soekarno, Effendy Sumardja, Michael Wells, Chris Wemmer, and Tony Whitten.

For various forms of assistance, ranging from identifying interviewees and providing contacts to offering hospitality during our travels, we thank Philip Rehbock, Harry Scheiber, Jim Findley, and Paul and Maryann Murphy. For granting us permission to use and modify plate tectonic maps, we thank Lawrence Lawver and Tung-Yi Lee and the Plates Project at the University of Texas at Austin.

This book owes a great deal to the photographers who helped us illustrate the beauty and diversity of Indonesia's life forms and landscapes. We are especially grateful to the following people for personally spending time with us to select luminous shots from their impressive portfolios: Alain Compost, Tim Laman, Greg MacGillivray, Kal Muller, Nick Nichols, and Merlin Tuttle. Jez O'Hare deserves our deepest thanks and admiration for taking on an arduous photo assignment specifically for this book, traveling to the remotest islands of the archipelago to capture on film what Wallace described in words.

Friends and family provided us with invaluable encouragement. We thank Chuck Cook, Dana and Taylor Cook, Carolyn Kato Daws, Kenji and Maruka Fujita, Bob and Peggy Barry, Dianne Dumanoski, Maureen Smith, Karin Johnson, David Rick, Tim Jessup, John McGlynn, Rick Pollard and Gadis Arivia, Ron Petocz, and Nengah Wirawan. We are grateful for the camaraderie and professional expertise of our editors at the University of California Press, Howard Boyer, Rose Vekony, and Anne Canright; our book designer, Steve Renick; and our photo editor, Laurel Anderson.

Lastly, a respectful salute across the years to Alfred Russel Wallace, whose profound, insightful, and superbly written book, *The Malay Archipelago*, invited us to see a land of beauty and natural riches through his eyes.

Note

In chapters 1 through 7 we have used the historical names and spellings for locations as Wallace gave them in *The Malay Archipelago,* including Celebes (now Sulawesi), the Moluccas (Maluku), Lombock (Lombok), Ké (Kei), Waigiou (Waigeo), and Amboyna (Ambon). Likewise, we have retained English measurements in chapters 1 through 7. In chapter 8 and in the captions and sidebars that deal with modern topics and present-day issues, we have switched to current political names and spellings for locations and to metric measurements. To aid the reader and provide a comparative reference, the map on pages 38–39, which traces Wallace's voyage through the archipelago, includes most of the historical names used in the text; that on pages 218–19, illustrating the locations of Indonesia's parks, uses modern names.

An appendix lists common and scientific names of species mentioned in the text, sidebars, and captions. Species names used by Wallace often differ from the current names and have not been changed in quotations. Throughout the text, Wallace's *Malay Archipelago* is used as a prime source, and his working is frequently paraphrased. Direct quotations are referenced in the endnotes.

PROLOGUE

Edward O. Wilson

At the dawn of field biology, Alfred Russel Wallace departed for the most distant and dangerous biotic frontiers of the world, carrying with him little formal education but a blessed love of reading and reflective solitude. He sought the insect-ridden Edens of which naturalist explorers dream. His principal lifeline to the English homeland consisted of specimens outbound—birds skinned, insects pinned, plants pressed—and sporadic payments for his treasures inbound. An intense young man, totally focused, awesomely persistent and resourceful, resilient to tropical diseases that killed so many others, and nobly selfless, even to Darwin, who otherwise might have become a bitter rival, Wallace endured, and he triumphed.

He succeeded brilliantly because he relished detail while thinking across a wide canvas. And he took aim at what I like to call the "Big Tropics": the Amazonian and Malaysian wildernesses, which harbor the richest faunas and floras in the world. With Humboldt, Bates, and a small company of others, he was the first to plumb their biotic depths. He did not just visit; he often lived as a resident in localities where almost every species was new to Westerners, even the most spectacular birds and butterflies, and important discoveries could be made with the ease of picking tropical fruit off the ground.

The greatest discovery of all, evolution by natural selection, awaited experience and imagination in the tropical archipelagoes he visited. Wallace, the autodidact, and Darwin, better trained but also an autodidact in those parts of science that mattered, were both perfectly positioned to make the discovery. It came to them independently as they traveled, with their observation that a species on one island is often matched by distant yet closely similar species on nearby islands. Each man asked, Might these vicariant species have evolved from a common ancestor? And if so, how? The answer that came to both was: Species arise by constant competition among varieties occurring in the same population. Competition is inevitable, because as Malthus had noted earlier for humans, population growth always exceeds the growth of available resources. Add the random production of new varieties, however minor, and evolution will proceed indefinitely and cumulatively up to and including the creation of novel species. Darwin received his insight from the birds of the biotically impoverished Galápagos Islands. Wallace got his from the butterflies and other organisms of the vastly larger and richer islands of present-day Malaysia and Indonesia.

The vastness of the tropical archipelagoes also provided the knowledge Wallace needed to conceive the biological discipline of biogeography, which has expanded during the late twentieth century into a cornerstone of ecology and conservation biology.

In *Archipelago,* Gavan Daws and Marty Fujita provide a brief but excellent biography of Alfred Russel Wallace. Their prose, featuring short, propulsive sentences and surprising factual twists, draws the reader paragraph by paragraph—as across a string of fascinating islands—through the life of their subject. But their goal is much more than a biographic sketch. They mean for the reader to glimpse the natural environment of Wallace's archipelago as he saw it, a spectacle hundreds of millions of years in the making, rich beyond imagination. It is still there, one of humanity's great endowments, but shrunken, fragmented, and under furious assault from both within and outside the countries that own it. To save its remnants for generations to come, someday to take the full measure of its glory, deserves the highest priority in global conservation.

ARCHIPELAGO

A frog basks in a sunspot in the rain forest understory. (Greg MacGillivray Films)

Painting of Alfred Russel Wallace by J. W. Beaufort, 1923. (Courtesy of The Natural History Museum, London)

THE EVOLUTION OF A NATURALIST I

Examine the infant Alfred Russel Wallace as a biological specimen, a form of life on earth, and the odds against his survival would appear statistically daunting. He came into the world on January 8, 1823, the eighth of nine children born to one of those alarmingly large nineteenth-century English families in which the death rate of offspring in their early years was close to fifty percent, and in maturity not much better. Of the five Wallace daughters, three died as children and one in her early twenties. Of the four sons, one died before reaching full adulthood, another before middle age. Alfred was the third of the four boys. In his schoolboy years he was ill a number of times, at least once to near-death. But he beat the odds and lived to be ninety.

His father was a middle-class man but not a reliable middle-class provider, no good with money, not a dependable worker, a dabbler. The only thing he generated steadily was babies. From such a father, Alfred said, the Wallace children inherited laziness. Yet Alfred's own working life yielded more than six decades of nonstop productivity of the highest quality.

As a child he did not appear well equipped for sustained, determined struggle toward eminence of any kind. He did not develop rapidly, and he was not greatly fond of school; he said he found geography as uninteresting as the learning of the multiplication table, in the painfulness of the process and the permanence of the result. He did enjoy reading—travel books, history, biographies—but that was outside class hours, by himself. Whatever intelligence and intellectual appetite he might have shown, there was no chance that he would go on to university. In his father's house there was no money for that. His formal education ended at fourteen.

The birthplace of Alfred Russel Wallace at Kensington Cottage in Usk, Monmouthshire, Wales. (Courtesy of Hunt Institute for Botanical Documentation / Carnegie Mellon University)

The working world of Britain in the late 1830s and 1840s did not appear to offer him a secure place. He spent some time in London with his older brother John, who was apprenticed to a builder, and then went to the country to join his eldest brother, William, a surveyor. But his own employment was never much more than marginal and sporadic, with no prospects for advancement.

Through his teenage years he was still, so to speak, in a chrysalid state, immature, only partially formed, and he remained outwardly quite unremarkable into his twenties. In his own homemade way he was a thinker, but thinking did not pay. At twenty-one, he said, he had never had a gold sovereign of his own.

He did not really try his wings—or even find out that he had wings—until he was twenty-five and a world away from England, on the other side of the Atlantic, in the tropical forests of the Amazon. There he began to live as an adult. There for the first time he engaged himself fully in his true life's work, as a naturalist.

He came back to England at twenty-nine, and at thirty-one was gone again, to the far side of the globe, to the Malay Archipelago. He spent eight years in the islands. He called the sojourn "the central and controlling incident" of his existence. This was when his life became truly his own and became truly remarkable.

Before he returned to England at thirty-nine, he had turned himself into one of the greatest field naturalists of his century. There was no more skilled observer and describer of living things. He was also the greatest collector of his time: if all the specimens he shipped back to London had been housed under one roof, they would have filled a natural history museum bigger than many in England. And in his scientific solitude in the islands he developed into a great thinker, a theorist on a world scale, as formidable as any on earth.

In the Malay Archipelago, Wallace had no professional associates. He could not attend the meetings of learned societies. He had no access to libraries or natural history museums. His only reference books were the few he could pack and carry with him. Yet with nothing more than the power of his own mind to draw upon, he wrote and sent back to England for publication more than forty scientific communications, on the widest possible spread of subjects under the broad heading of natural history, ranging from brief to lengthy, from particular to general, from useful to seminal to revolutionary.

One of these communications, drafted on flimsy paper in the solitude of one of the most remote islands of a remote archipelago, was transformational on a world scale. In less than four thousand words Wallace laid out the principle of the evolution of species by natural selection, which became the new controlling idea of biological science, and above and beyond that one of the most influential ideas in the history of all of Western science—indeed, of Western culture at large.

Upon this high mountain peak of human thought, Alfred Russel Wallace stands with Charles Darwin.

The metamorphosis of Wallace in his later twenties and thirties is an arresting phenomenon. Toward the end of his life he looked back on himself as he had been at twenty-one, and what he saw in retrospect was less than encouraging. He had little more than the workingman's clothes he stood up in and enough to buy an occasional book. He did not drink much, and he smoked not at all. He had no ear for languages or music and no real eye for art. He did not cut a social dash. He was not much of a talker in company. He liked to listen to a good storyteller, but did not think he was one himself; the spoken word came to him too slowly. He did consider that with time to reflect privately and pick his words, he might perhaps have it in him to write. He was tall—six feet one—but his body was not an asset. He was good with his hands, but his nervous system was delicate. He had to wear eyeglasses. His lungs were not strong. When he was quite young he nearly drowned in a river near his home; he did not learn to swim until he was almost out of his teens, and even then he was no good at it. After twenty yards he could not keep his mouth above water; his legs, unusually long and heavy, pulled him down, and being what he called rather deficient in strength of muscle, he became disinclined to practice what he felt to be beyond his aquatic powers. Summing himself up physically and mentally, he saw constitutional laziness and a general disinclination to exertion—what he called a want of courage and assertiveness.

Wallace's self-diagnosis was validated and elaborated for him by phrenology, a popular science in his time. The theory was that the contours of the human skull could be read as a kind of relief map of the mind, so that mental attributes could be deduced from the topography of the head. The practice was taken seriously by many, all the way to the highest levels of society: phrenology was suggested as a scientific method of choosing better-quality members of Parliament, and the royal family brought in a phrenologist to look into the educational retardation of the Prince of Wales.

Wallace had the bumps on his own skull read, twice, and he came away a believer. His want of *self-confidence* was palpable. His bump of *individuality* was small, and it was true that he was not good at recognizing and identifying other people. He had better bumps for *locality* and *color*, and indeed he could always recall physical reality, especially outdoors, vividly and in detail, all the way back to his earliest childhood. *Inhabitiveness,* attachment to place, was small; *desire to travel* was large. He also had significant bumps for *comparison, causality, order, firmness, acquisitiveness, concentrativeness,* and *constructiveness.* His bump of *veneration* was small, which correlated with the fact that he was not conventionally religious. But *ideality* and *wonder* were large, signifying an intense delight in the grand, the beautiful, and the mysterious, especially in nature.

This was Alfred Russel Wallace at twenty-one: a tall, gangly young man, intelligent in his way, thoughtful, sensitive, but at the same time shy, reticent, awkward to the point of being uneasy about sociability, happiest alone. He had no professional qualifications and no aptitude for moneymaking, in fact a revulsion against the very thought. If he had to work, and of course he did, then his taste—his need—for solitude meant that surveying suited him as well as anything could have: quiet days following the chain, breathing fresh air, lunching on bread and cheese by a rippling brook, with the countryside spread out before him for nature rambling on his days off.

Wallace's nature rambles started as nothing more than Sunday strolls. He did not know there was such a thing as a science of botany. He was not even familiar with the common names of trees, shrubs, grasses, and flowers. But he responded to "the influence of nature," and this turned into a positive appetite. He collected plants. He bought a book on the elements of botany. It cost him ten shillings and sixpence—for him a lot of money, just over half the gold sovereign he never had in those years. He built on what the book told him by copying information from an encyclopedia about British botanical species. The more he studied the parts and organs of flowers, the more specimens he added to his herbarium, his collection of pressed and dried plants, the more enthusiastic he became to learn still more.

Wallace walking in the woods was a solitary figure in a crowded country landscape. Britain was full of nature ramblers. To enjoy nature, to observe, collect, classify, and arrange "natural productions"—this was a flourishing national pastime. A country squire, an aristocratic shooter of game, might also maintain a nature preserve on his estate; his wife and daughters might not care to follow the guns, but they might be bird-watchers. For comfortable upper-middle-class women, underoccupied beyond the responsibility of supervising servants, nature rambling was a tasteful way to consume spare time. In the opinion of the higher orders of society, nature rambling was even good for the lower classes: it kept them out of the pubs, if only for the time of the ramble; and if they congregated back at the bar to discuss what they had found, at least the talk would be improving. All in all, nature rambling was seen as good for everyone, body, mind, and soul. It had the blessing of the clergy. It was Victorian.

Botanizing, gathering ferns and forest mosses, turning over hollow logs for beetles, netting butterflies, fossil hunting on chalk cliffs with a geologist's hammer, taking trips to the beach to look for seashells on the seashore, dredging for mollusks in the shallows, wading in a tide pool to scoop up little marine life-forms and put them in a jam jar—collectors were everywhere, from the beach at Brighton to the Lake District to the top of Mount Snowdon. The range of rambling was extended by roads paved with macadam, and even more by the coming of the railway. (Most of Wallace's surveying work was for new railway tracks, and it was while he was surveying that he became a collector.) Nature excursions by rail were very popular; plant spotting from the window of a train carriage was exciting. At home the pleasures were tranquil—watching the fish in the aquarium, or gathering as a family around the microscope, a window that opened onto another dimension of the natural world, teeming with life.

There were guidebooks for the proper enjoyment of nature, magazines devoted to natural history, and manuals of technical instruction: how to press and dry flowers, how to maintain caged birds in good health and spirits, how to preserve eggs, how to pin and mount insects for display, how to lay out seaweeds to the most pleasing artistic effect. And how to arrange a home cabinet of curiosities.

The cabinet was the physical container of nature enthusiasm. It might be a box, a shelf, a side table in the drawing room, or a glass-fronted display case. The bigger the cabinet the better, and the broader the range of specimens the better too. People exchanged local curiosities for distant ones. And for those who could afford it, London had a market in curiosities: ferns by the bushel basket from the most distant parts of

The leisure class of Victorian England, ever curious, loved to spend time browsing zoological collections in the British Museum. (L. Jewitt / The Mary Evans Picture Library)

Opposite: Scientists at work collecting and examining specimens aboard the HMS *Challenger* in 1874. (Mary Evans Picture Library)

England and exotic items from faraway places, sold at auction: South American orchids, seashells from Amboyna, monkey skulls from the African jungle, bright feathers of strange birds from tropical islands with unpronounceable names. In this way, private cabinets were connected with the wide world.

There were also big buildings and extensive open spaces that amounted to official British cabinets, civic and national collections of natural productions and curiosities drawn from all parts of the globe. With every decade, more of these museums, botanical gardens, and zoos were opened to the public. At midcentury, the Great Exhibition in the Crystal Palace was a huge cabinet of British industry under glass; a cabinet of science too, including life-size concrete dinosaurs.

And there were popular exhibits of natural curiosities, both homegrown and foreign—entertainments all the way to freak shows that charged admission to see the five-legged cow or the card-playing pig, the giant, the midget, the Siamese twins, or the latest imported "savages" being exhibited for the amazement and amusement of the civilized: Laplanders, Red Indians, Zulus, Pygmies, Aborigines, Aztecs.

European naval expeditions were out everywhere, doing hydrographic or geological surveys, and most ships carried naturalists of some description. A British expedition went as far as the far south of South America, the ends of the earth, and among the strangest things it brought back were specimens of the most extreme of savages, natives of Tierra del Fuego. One of these curiosities arrived dead in a barrel, pickled, for study at the Royal College of Surgeons. The living Fuegians had phrenology done on them, and ranked according to nineteenth-century European diagnostics of skull shape they turned out to have large bumps of *animality* and only small ones of *intellectuality*.

With Western nations establishing colonies in distant places and different climates, a global traffic in nature grew up. Some of the biggest, best, and most influential botanical gardens in the world were in the colonies. Plant transfer and economic botany were important: taking useful or profitable plants—hemp, rubber, cotton, tea, sugar cane, tobacco—from one colonial environment and establishing them in another for commercial harvesting. There was a strong interest in acclimatization; in the century up to 1850, about five thousand species of foreign plants were introduced into Britain. For the benefit of science, fossils from Africa and South America were shipped to London for study, also strange animal specimens from Australia—by 1834, five hundred platypuses, for dissection. The leading European naturalist of the middle and later eighteenth century, Linnaeus, had listed 444 known species of birds in 1758; another naturalist in 1817 listed 765. By the early 1860s the number was more than eight thousand and growing.

The store of information about nature was increasing by orders of magnitude, especially about plants. There even emerged a scientific specialty known as botanical arithmetic: counting the different species in a given area and determining the ratios between them. This was an attempt to impose meaning on factual overload by mapping nature in a new way.

The outlines of the great landmasses of the earth were well enough known, and wherever a region had been more or less systematically studied, maps could be drawn showing plant and animal life. Still, speaking botanically, and even more so zoologically, most of the earth outside Europe was unsurveyed by Europeans, its wealth of natural productions unknown, uncollected, unarranged, unclassified, uninterpreted: much of the western part of North America, most of South America, most of Africa, Australia, India, China—and not just the large landmasses but also the islands and the oceans of the world.

The piling up of natural facts, and the mapping of those facts for meaning, were certain to continue through the nineteenth century. The British vision was of empire and science advancing hand in hand. The scientific side of this grand project was to collect and bring home every species on earth, from insect to oyster to orchid to elephant.

The window of opportunity was huge, and the intellectual challenge enormous. How was all this information about nature to be understood, interpreted, and controlled?

New learned societies were formed. The Linnean Society, dating from 1788, was named for the preeminent scientist who developed a globally applicable system for classifying and naming plants and animals. The Geological Society was begun in 1807, the Zoological Society in 1826, the Royal Geographical Society in 1830, the Entomological Society in 1833; and more followed.

At society meetings and at universities, the broadest and deepest questions about nature were being debated. Most pressingly, if new information conflicted with old understandings, what then? In particular, what if new information about nature called the Bible into question?

As early as the seventeenth century, Noah's Ark was beginning to look impossibly overcrowded. The story of the Flood, and of Mount Ararat as a single place of dispersal from which all the saved animals went out two by two to multiply and populate the earth, grew more and more difficult to sustain as literal truth.

One way to handle questions raised by natural science without having to answer them was to say that the world of nature was organized according to God's will. God made nature bountiful and beautiful. He created species of animals, birds, and plants, separate and distinct, each species to live in its own place according to its own nature, unchanging, presumably for eternity. In the contemplation of nature, humans could see the power and the glory of God. That was what nature was for. If there were mysteries (why was the caterpillar created, and created so ugly? the peacock's tail was beautiful, but what was the use of it?), then that was God's will too.

This was established religion speaking. And there were many, not just clergymen and good Christian laymen, but orthodox scientific thinkers as well, who did not want to see established beliefs challenged and established faith undermined. But now established faith and established understandings of nature, and perhaps even established society itself, were under threat from all this new information in natural science. The ground was shifting underfoot. Where to stand?

While Wallace was growing up, huge questions were being wrestled with concerning the earth itself. How old, really, was it? Thousands of years? Or millions? Across millennia and eras and eons, how had it altered? On the presently visible earthly evidence, there had been ice ages in the past. What caused radical global climatic change? Mountain ranges had risen, sea levels had fallen and risen again, meaning that the land area of continents had expanded and shrunk, islands had disappeared and reappeared. What were the mechanisms driving such huge geological convulsions?

Had the changes of the ancient past been sudden and catastrophic? Or was change over long stretches of time steady, incremental, "uniformitarian"? Were the climatic and geological mechanisms at work in the present—wind systems, ocean currents, glacial action, erosion, volcanoes, earthquakes—the same as those that had operated in the past? Overall, could scientific knowledge of the present condition of the physical earth explain its past? And vice versa?

And then, what might geological and climatic change have to do with the understanding of natural life? Why were there so many species of animals, birds, insects, and plants on earth, with more being discovered all the time, and no end in sight? Was it possible to connect fossil species with present-day species? What was the significance of variations within species? Where, precisely, was the scientific borderline between a variety and a species? Why were species distributed geographically the way they were? Why were there distinct regional aggregations of species? What was the cause of geographical continuity or discontinuity in these distributions?

And the ultimate question: What was the connection of all of this with human life?

No thinking person could ignore the growing tensions between old understandings and new information. And no one was more acutely and uncomfortably aware of the subversiveness of new ways of looking at nature than Charles Darwin.

A photograph of Darwin ca. 1854.
(Maull & Fox / Mary Evans Picture
Library)

Charles Darwin.
Cir–1854.

Darwin was born into the comfortable established orders of British society. His family had plenty of money, and he married more. He went to university at Cambridge and was intended to be either a doctor like his father or a clergyman of the established Church of England. He became neither. Growing up, he was a nature rambler, a collector of beetles, a shooter of birds. At twenty-two, courtesy of social connections—not professional scientific standing, of which he had none—he was given the chance to travel on a round-the-world naval expedition as a companion to the commander (the same man, as it happened, who on an earlier voyage had brought the Fuegian curiosities to England). What Darwin saw during the voyage of HMS *Beagle* (1831–1836), and what he concluded from everything he observed and collected, converted him to materialist science and committed him to the working out of revolutionary biological theories that absorbed and obsessed him for the rest of his life.

For many years he could not bring himself to publish what he was privately convinced was the scientific truth, because it would be so destructive of old social and religious understandings and agreements. The new order coming to life in Darwin's ideas was death to the old order.

Wallace was younger than Darwin, and he was born into a less elevated reach of society. He grew up far less beholden to the old order of things. He was not from a privileged family with inherited money. He never went to university. And he never assumed that God was an Anglican and that all was right with His world. The human life on earth that Wallace saw around him was riddled with inequities and injustices, and if this was God's will, then Wallace did not think much of God. When he read the Bible, he found large stretches of it simply unbelievable.

From age fourteen Wallace had to live a working life, and among working people he came into contact with any number of thoughtful men who had never been to Oxford or Cambridge, who did their reading at public libraries, their talking about the nature of life at mechanics' institutes. Some were religious dissidents or freethinkers, some were socialists, some were students of natural science. Wallace in his own chrysalid way was all of the above.

Books focused his nature rambling and shaped his evolving ideas about science. Between his middle teens and his early twenties he graduated in his reading from shilling publications of the Society for the Diffusion of Useful Knowledge, through botanical manuals and encyclopedias, to original and mind-extending works by highly intelligent scientific travelers—Darwin's *Journal of Researches into the Geology and Natural History of the Various Countries Visited by H.M.S. "Beagle"* and Alexander Friedrich von Humboldt's *Personal Narrative of Travels to the Equinoctial Regions of the New Continent* (meaning South America)—to serious scientific compendiums such as James Cowles Prichard's *Researches into the Physical History of Mankind*, to the most wide-ranging and deeply thought-out synthesis of information and theory about life on earth, Charles Lyell's *Principles of Geology.*

The year 1844 was a marker in Wallace's development. His surveying work had dried up, and he was lucky to find a modest teaching job in Leicester. The town had a good library, and one of the books he read there was Thomas Robert Malthus's *Essay on the Principle of Population.* Malthus argued that human populations had the biological potential to multiply indefinitely—exponentially—whereas the resources to sustain them, especially food, could only increase arithmetically. So, by the very nature of life on earth, the growth of human population was bound to be checked, in the extreme by starvation and disease, even war. This was a grim vision of life as struggle. Malthus's ideas struck Wallace with great force. They became what he called a "permanent possession."

Another important encounter for Wallace took place in the library at Leicester when he met Henry Walter Bates. Wallace and Bates were two of a kind—shy young men of no social standing with a strong appetite for amateur natural history collecting. Wallace was a botanizer, Bates went beetling. They became friends. When Bates showed Wallace his beetle collection, Wallace was amazed at "the great number and variety . . . and more amazed to find that they had been collected around Leicester and there were still many more to be discovered." He bought himself a collecting bottle, pins and a store-box, and a manual of the British coleoptera. After he left Leicester to go to work again with his brother John, he and Bates kept in touch, writing letters back and forth about insects and ideas.

In 1847 they spent some days beetling together, and the notion came up of an overseas collecting expedition. Wallace called it "rather a wild scheme." That it was. He and Bates between them had next to no money. Neither did they have the connections or the credential for a naturalist's berth on an official ship. Whatever they might manage to do, it would be entirely on their own account, and speculative. They would have to pay their way by selling specimens of what they collected.

Where to collect? Unhesitatingly, they picked the tropics. From the first time Wallace saw pictures of orchids in a nurseryman's catalog, he had felt a pull in that direction. Exactly where? A book called *A Voyage up the Amazon* had just come out, written by an American named William Edwards. It made the region sound mysterious and at the same time easy to get to. And it seemed to hold out the possibility of earning their way by "making collections in Natural History."

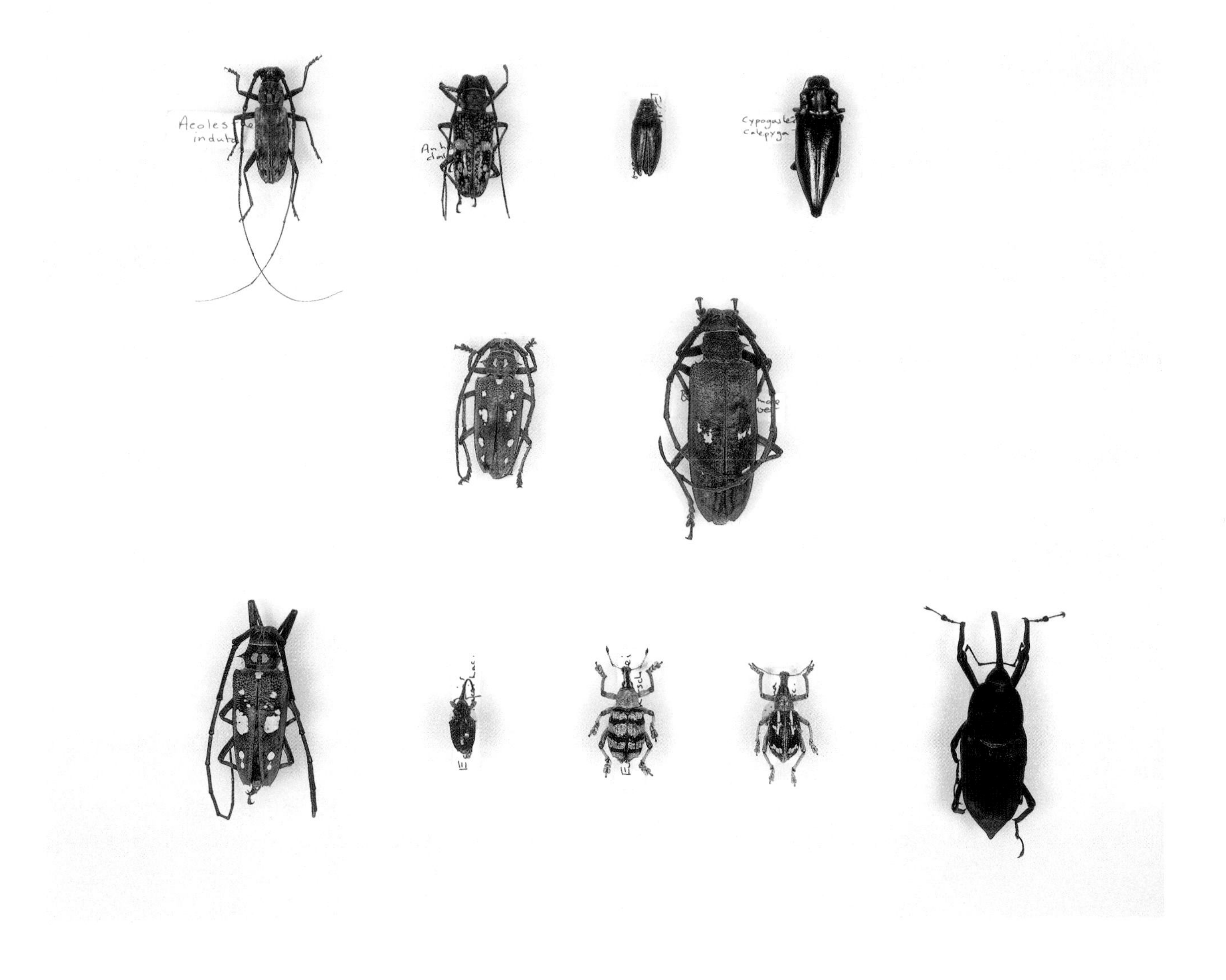

One of Wallace's early collections of beetles. (Courtesy of The Natural History Museum, London)

Wallace and Bates were not the first British traveling collectors to point themselves at South America. Charles Waterton, for one, had been there, over twenty years before. Waterton was a Yorkshire squire, a gentleman with a nature reserve. He could pay for his excursions out of his inherited wealth. On top of money to spend however he wanted, he had a wild sense of humor all his own. He brought back from the jungle a head that looked equal parts human and nonhuman. He called it the Nondescript. It was actually a red howler monkey head that he had doctored—he was a very skillful skinner and preserver. The other Watertonian claim to fame was that on the Demerara River he had ridden a twelve-foot caiman, a South American alligator, wrestled it onto a sandbank, cut its throat, and then sat down to eat breakfast before skinning it. At home he kept the caiman in his house, stuffed. And he liked to lurk under a table in the hall and bite visitors in the leg.

Wallace and Bates were not Watertonian wild men. They were as serious and sober as they were young. The book that attracted them to the Amazon was not Waterton's but William Edwards's. They had other, more substantial, books in mind as well: Darwin's *Journal*, Humboldt's *Travels*, and the anonymously published *Vestiges of the Natural History of Creation*, which was full of speculation about the way species changed over time. Their plan was to test deep ideas and theories of natural history against the facts of natural life, make scientific collections, and stay alive by selling duplicate specimens.

In London they found an agent, Samuel Stevens, to take delivery of specimens and put them up for sale. William Edwards happened to be in town; they met him, and he wrote letters of introduction for them. They went to see the palm trees and orchids at Chatsworth House. They looked at a butterfly collection from the Amazon. They talked to Thomas Horsfield, who was running the British East India Company's museum. Horsfield had been in the tropics—not in the Amazon but in Java, where he lived in a house full of natural curiosities. He had done a botany of Java and had sent plants back to the botanical gardens at Kew. He had collected butterflies, and he showed Wallace and Bates the kind of wooden case he used for storing and shipping specimens. They were going to be collecting birds as well, a bloodier business than pinning insects and pressing dried plants. They spent a week practicing shooting and skinning. They did not visit Charles Waterton.

Wallace had himself vaccinated and bought some spare eyeglasses. On April 20, 1848, he and Bates boarded their ship, a merchantman, at Liverpool. They were the only passengers. Wallace was seasick for five days.

Wallace spent his next four years in the tropical forests of the New World, living a life completely new to him, strange, absorbing, demanding. He shot whitewater rapids in a canoe—he, the seasickness-prone next-to-nonswimmer, fearful as a child about crossing a river on a ferry boat. With his weak myopic eyes, he read a newspaper at night by the light of fireflies. He ate what the jungle set before him, declining chameleon and odd-looking frog, accepting alligator tail on a stick, turtle egg omelette, fried monkey, and great red-headed ant, preferably fried or smoked, with a sprinkling of salt. Amazon ants were everywhere: big ones as much as an inch and a half long, smaller ones but with a big bite; ants that ate collected specimens hung out to dry, ants that could work their way into all but the most tightly sealed boxes, armies of ants that formed chains to carry away food, making off with the farinha grain by grain, so persistent and unstoppable that Bates took to laying gunpowder along their trails to blow them up.

The young, socially uncertain Wallace was a persevering and open-minded traveler whose peculiar work aroused the curiosity of the native South Americans. Note the eyeglasses: his sight was weak. (Courtesy of the Hunt Institute for Botanical Documentation / Carnegie Mellon University)

On boring evenings Wallace wrote blank verse. On less boring evenings he watched Indian women dancing, and appreciated what he called the "perfect nudity of these daughters of the forest." He came across other Indians armed with blowpipes for shooting poison darts; white river traders given over to drinking and gambling; a man living deep in the jungle with a harem—a mother, her daughter, and two Indian girls, all of the females working away, turning out wonderful featherwork; a debauched old priest with racy stories about nuns; a half-breed murderer living in exile. He witnessed slavery and did not like what he saw, not even when the slave master was benign.

Wallace and Bates made a friendly decision to go their separate ways, Bates to the upper Amazon, Wallace along the Rio Negro as far as the Orinoco, which was Alexander von Humboldt territory. He was determined, he said, to see and know more of the place and the people than any other European; and if he did not get profit, he hoped at least for some credit as an industrious and persevering traveler.

Wallace the solitary, the socially uncertain, had to learn to negotiate with people of the towns, merchants and bureaucrats, and, far more importantly, day by day, with people of the jungle, hunters and gatherers.

He was open to the idea that native peoples were worth something. Not all Westerners of his time were so minded, but Wallace for his part could come across what his century called a Savage and be impressed by what he saw as Man in the State of Nature, walking with the free step of the independent forest dweller, original and self-sustaining as the animals around him.

At the same time, Wallace was Civilized Western Man, in the forest on business. He needed carriers and paddlers and hunters to work for him; but the Savage might have ideas entirely different from Wallace's about the value of time and money, about obligation—and beyond that, entirely other conceptions of the world and of Wallace's place in it.

A toucan, one of the many exquisite birds that Wallace encountered in his travels in the Amazon. Painting by Nicholas Aylward Vigors, 1831. (Courtesy of the Natural History Museum, London)

White men were strange anyway, and Wallace was very strange, catching and killing living things and putting the dead bodies in boxes. Skinning and stuffing, Wallace drew crowds, and his bizarre occupation was the subject of much discussion. It was decided that the markings of butterflies were for new patterns for printed calico. Ugly insects were for making medicine. But birds, for show? The English could hardly be such fools as to want to see a few parrot and pigeon skins.

In all of England there were seventy species of butterflies. In the Amazon, four hundred species could be collected in the first two months. The forest was a natural orchid house; Wallace once collected thirty species in an hour. And the place was prodigal in animals: all kinds of monkeys, all kinds of birds, snakes big and small.

Wallace shot at everything that crossed his sights. He skinned alligators and stuffed turtles. He collected fish that the Indians killed with poison. And he saw a black jaguar. It came out of the jungle twenty yards in front of him and looked him in the eye. Wallace felt no fear. He had his gun but did not shoot. Partly from surprise, partly from prudence—he had both barrels loaded, but only with small shot. And partly from ad-

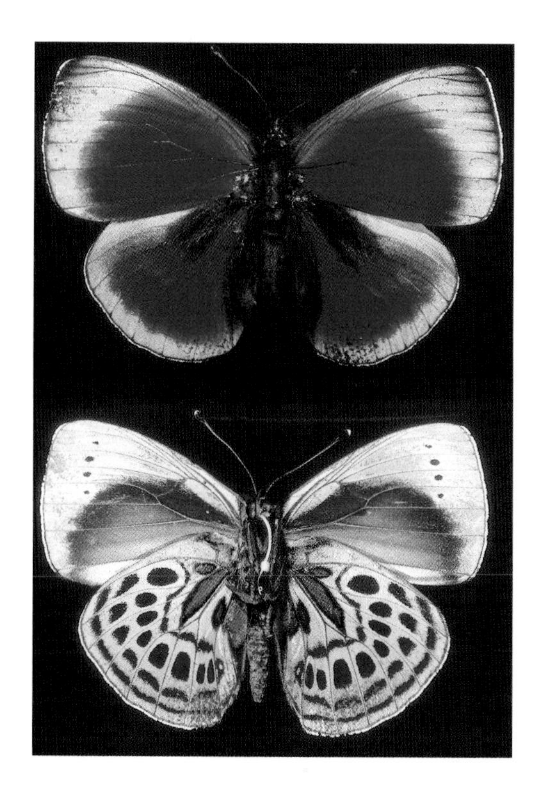

An example of the beautiful butterfly fauna of the Brazilian Amazon, *Callithea leprevri.* The photograph shows both sides of the specimen. (Gary Retherford / Photo Researchers, Inc.)

miration: "I had at length a full view, in his native wilds, of the rarest variety of the most powerful and dangerous animal inhabiting the American continent." The jaguar went its way, disappearing into a thicket, and Wallace could hear small animals and ground birds scampering and whizzing off ahead of it.

Once Wallace accidentally shot himself in the hand. The wound became inflamed, and he spent weeks with his arm in a sling, miserable, unable even to pin an insect. He was attacked by ants, gnats, fleas, sand flies, chiggers, swarms of wasps, mosquitoes, vampire bats that sucked his blood in his sleep, and the *pium,* an insect whose bites in the hundreds raised bloody spots and left his feet red and swollen. Any number of times he was ill, racked by malarial fever, dosing himself with quinine, out of action for days and weeks at a stretch.

His younger brother Herbert, the last of the nine Wallace children, came out and joined him. He stayed a year—long enough for both of them to realize that he was not up to the demands of the collecting life—and decided to go home, got as far as the coast at Pará, caught yellow fever, and died.

Wallace was a survivor of the jungle, and more. Along the Amazon and the Rio Negro he labored nonstop, collecting specimens, accumulating detailed information, expanding the territory of his knowledge, making sketches and maps, forever trying to construct patterned meaning out of his gatherings. Different varieties of fish in different tributaries of a single river; one kind of monkey on the near side, another on the far side: geography was crucially important to his sense of nature. It was what determined the distribution, the range, the habitat of living things—the *where* of life.

Wallace was an enthusiast for the tropics. He had a vision of an endlessly fruitful earth, where a minimum of labor could transform the forest into a humanscape of beauty and variety, with the prospect of a glorious existence of ease and plenty, free from the money-matter cares and annoyances of civilization.

This version of the tropical life was a recurrent fantasy of cold-weather, industrial-society Europeans. Wallace put it on paper, but in the working life he was actually leading along the Amazon and the Rio Negro, it was really no more than a daydream. He labored to exhaustion and beyond.

Now and then, and especially when he was ill and bone-weary, he felt a pull toward home. After four years he decided it was time to go.

Headed downriver toward the coast for the last time, he was bringing with him fifty-two live animals. By Barra, the number was down to thirty-four—five monkeys, twenty parrots and parroquets of twelve different species, two macaws, a white-breasted Brazilian pheasant, five smaller birds, and a toucan, fully grown and very tame; he had great hopes of bringing it alive to England. By Pará the numbers were down again. A little black monkey had killed and eaten two of the parrots, several more had died, and the toucan had escaped.

All along, Wallace had been shipping boxes of specimens from the interior to the coast, consigned to London for Samuel Stevens to sell. At Pará he found box upon box, which through the workings and nonworkings of colonial bureaucracy had never been shipped. Packed inside were thousands of specimens, the yield of his years of hard labor, amounting to his business capital, his professional capital, his life capital.

He managed to get passage to England for himself and his boxes and his live birds and monkeys on the brig *Helen*. He went aboard ill with the ague. On the open ocean he was seasick. Then, three weeks out into the Atlantic, the ship caught fire and had to be abandoned.

Wallace went over the side and into a lifeboat with only his pocket watch, a tin box containing some shirts, drawings of fish and animals, and notes on palm trees. Through the day and into the night, he watched the ship burning. All his boxed specimens burned with it, all his monkeys, and all his parrots, except one that fell into the water and was picked up.

The *Helen*'s two little boats drifted ten days before they were sighted by a merchant-man. The *Jordeson* was a lifesaver; but that much having been said and appreciated, the ship was a miserable thing, capable of making only fifty miles a day in the direction of England. Along the way it ran into a storm—heavy seas that made it stagger like a drunken man, winds high enough to blow its sails away. When at last it reached the English Channel, another severe delaying storm came up. Wallace was eighty days out of Pará before he set foot on dry British soil with his tin box.

He regretted his money loss, but even more the loss of what would have been one of the finest collections of American species in Europe, and on top of everything else the loss of years of journals, written natural history observations, and sketches. He was, he said, in some need of philosophical resignation to bear his fate with patience and equanimity.

On shore at last, he rejoiced in beefsteak and damson pie. In less than a week he was in London at a meeting of the Entomological Society and his mind was turning to another collecting expedition. "Fifty times since I left Pará have I vowed, if I once reached England, never to trust myself more on the ocean, but good resolutions soon fade, and I am already only doubtful whether the Andes or the Philippines are to be the scene of my next wanderings."

While he was ruminating the question of where in the world to go he published two books, one a narrative of his naturalist traveler's tales, the other a technical work on palm trees. Neither made him any appreciable money. Through the months of writing and beyond he was able to eat only because his careful agent, Samuel Stevens, had taken out insurance for him against the loss of his specimens.

He had enough money for a new suit. He wore it to meetings of learned societies—the Linnean, the Entomological, the Zoological. He heard Thomas Huxley speak and was greatly impressed by the intellectual force of the man, his ability to make a complex subject intelligible and interesting. Wallace was too shy to introduce himself. In the jungle a black jaguar could fix him with its gaze and he felt no fear. But at home, confronted with high-powered humans, he was still Wallace the diffident. At the insect room of the British Museum he had met Charles Darwin and left no impression.

The Naturalist's Cabinet

Fig. 255.—The umbrella and its mode of use. *(After Kiesenwetter.)*

A whimsical means of collecting insects. (North Wind Picture Archives)

Collecting, classifying, and contemplating objects from nature was, for the Victorian British of Wallace's day, a popular pastime, an engrossing intellectual exercise, and a serious scientific endeavor. It drew in people from all social classes and was regularly featured in journals and magazines across the country. Contemporaries often remarked on this upsurge of interest in the natural world.

The nineteenth-century fascination with nature was not without precedent: it owed much to older traditions of natural history study and collecting. The first substantial collections of objects from the natural world were created in the seventeenth century by aristocrats, gentlemen, and would-be gentlemen. These individuals, with money and time to spare, began to collect objects, or "rarities," of art and nature while on holiday in Europe. At home, they displayed their finds in "cabinets of curiosity"—rooms set aside for the purpose. Displays were often wildly eclectic: fossils, coins, stuffed birds, pickled fish, even the putative feathers of mythical creatures. The aim was to inspire fascination with the rare, novel, and surprising in all spheres of God's work.

In a more scientific vein, the seventeenth century also saw the formation of the Royal Society in London. Members, unlike the self-styled "curiosi," saw themselves as men of learning. They regarded collecting not as a means of displaying wealth and taste but as a way of gathering useful information and developing a theoretical understanding of the natural world. The society accordingly issued instructions to travelers, sought information from corresponding members across England, and began systematic collections. Their aim was to discover regularities rather than rarities in nature; they insisted on collecting both the common and the exotic.

This systematic scientific approach to natural history was greatly advanced in the next century with the publication of Carl Linnaeus's *Systema Natura* in 1735. Subsequent editions, published in pocketbook form, were extended to complete classification for the animal and plant kingdoms. The great advantage of Linnaeus's system was that it was both all-encompassing and easy to apply. It could be used by both the amateur collector and the serious scholar. Life was divided into six levels of generality: class, order, family, genus, species, and variety. Species, defined as organisms that could reproduce, were classified based on their common characteristics.

Linnaeus also introduced the modern system of binomial nomenclature, which came to be used across Europe and then across the globe as European explorers brought in collections from overseas. Each organism was given two Latin names: the name of the genus followed by that of the species. As collecting proceeded, new species were generally named after the discoverer or the discoverer's wealthy patron. The system thus came to be a way of rewarding those on whom the naturalists depended for information or support.

With the popularization of the Linnean system, the search for ever more complete collections of plants and animals was begun in earnest. Networks of institutions in Britain and Europe were established to organize and manage the work of collecting. In 1784 Britain acquired Linnaeus's own precisely cataloged collection, and in 1788 the Linnean Society was formed. Amateur collectors—country parsons, artisans, mechanics, schoolteachers, and surveyors—joined the activity, gathering together in field naturalist clubs and mechanics' institutes to listen to lectures and organize collecting excursions. Enthusi-

One of Wallace's many display boxes showing off his rich butterfly collection. (Courtesy of The Natural History Museum, London)

asts wrote and popularized natural history manuals. In towns and cities, naturalists and civic-minded citizens established museums, zoological gardens, and botanical gardens to store and display collections from Britain and overseas. In conjunction with the museums, scholars and amateurs formed scientific societies to advance the increasingly specialized work of gathering information and theorizing about the workings of the natural world. With this development, collecting became professionalized. For the first time men like Wallace could earn their living by collecting natural history specimens.

To meet this nineteenth-century collecting boom, a wide array of field equipment was devised. Botany boxes or sandwich boxes for plants were produced in all sizes and shapes. Specialized knives for cutting and trowels for digging became standard equipment. For butterfly enthusiasts, all sorts of nets were manufactured. Preservation was equally important. For insects, there were jars and pinning boards; for plants and ferns, the Wardian glass-enclosed case, which preserved plants by trapping the moisture they transpired. For home collections, whether composed of shells, beetles, seaweeds, or feathers, the personal cabinet, typically a wood-and-glass case, became a standard item in middle-class houses.

The popularization and professionalization of collecting did not mean that the attitude of wonder first cultivated by the curiosi of the seventeenth century was lost. On the contrary, nineteenth-century Victorians of all classes continued to be fascinated by nature's marvels. They combined an interest in the careful Linnean classification and labeling of specimens with an aesthetic delight in the delicate forms of plants and the intricate details of living creatures. Moreover, their fascination with nature broadened the audience for more theoretical concerns having to do with natural history, such as the origins of the species they so admired.

Sturdy wooden Bugis schooners still sail the seas of the archipelago. Wallace depended on many types of local sailing vessels during his eight-year adventure. (Jez O'Hare)

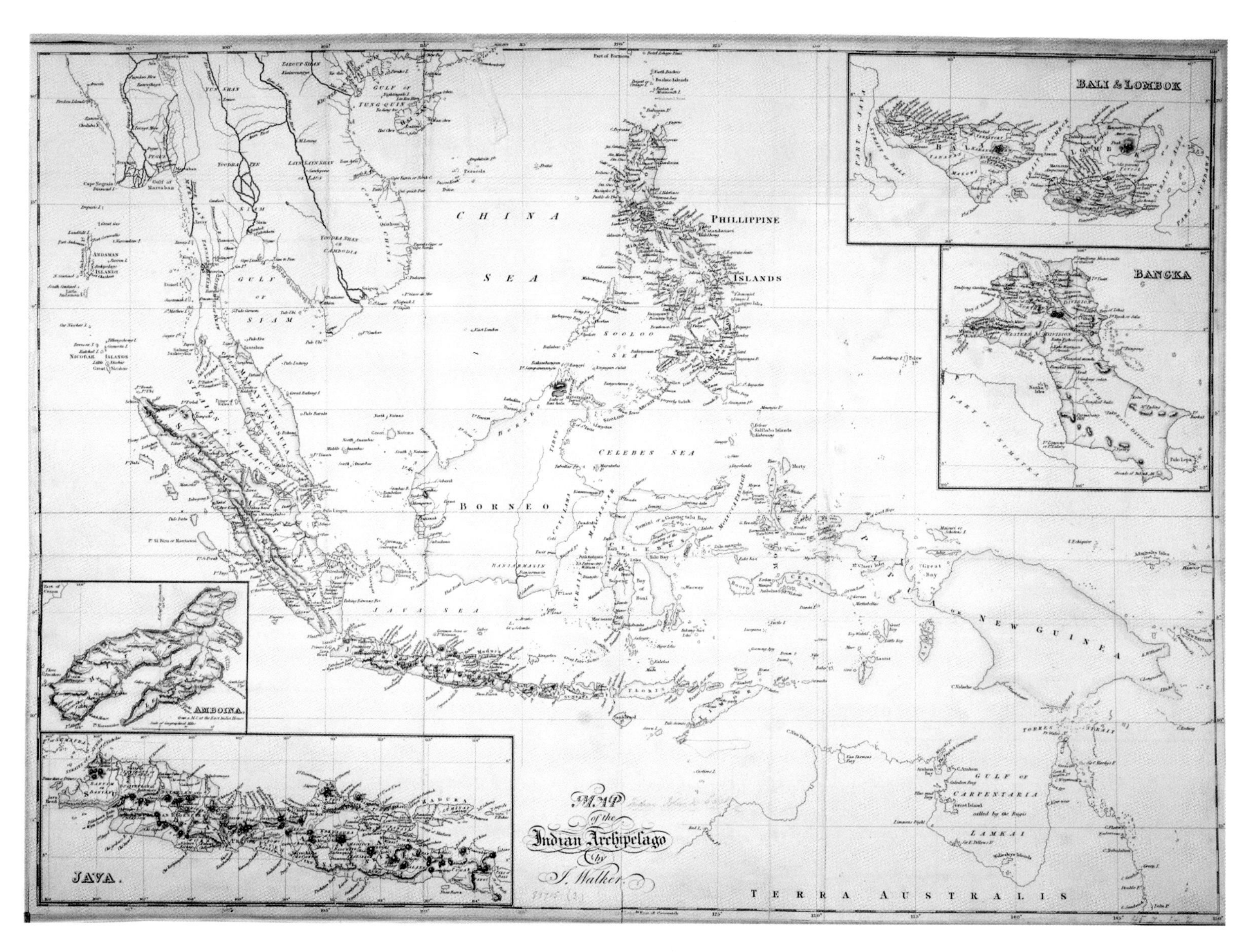

A map of Southeast Asia drawn in 1815. As a region, the Malay Archipelago was as large as the other primary divisions of the earth. (Courtesy of the Map Room / The British Museum, London)

2 PREPARING FOR THE ARCHIPELAGO

In the insect room, as everywhere else, Wallace was doing reconnaissance for his next expedition. Where to: the Andes? the Philippines? Africa? He decided on the Malay Archipelago.

Enough was known—and not known—about this part of the world to make it the place for him. In the words of his own *tour d'horizon,* "If we look at a globe or a map of the Eastern hemisphere, we shall perceive between Asia and Australia a number of large and small islands, forming a connected group distinct from those great masses of land, and having little connexion with either of them." As a region, the archipelago was comparable to the other primary divisions of the earth. It stretched four thousand miles east to west and thirteen hundred miles north to south across the equator. The biggest islands, Borneo, Sumatra, and New Guinea, were each as big as the British Isles or bigger; Java and Celebes were each the size of Ireland. But "to the ordinary Englishman this is perhaps the least known part of the globe. Our possessions in it are few and scanty; scarcely any of our travellers go to explore it; and in many collections of maps it is almost ignored." Yet it was full of striking natural phenomena—"one of the chief volcanic belts upon the globe," for example—and natural opulence: "Bathed by the tepid water of the great tropical oceans, this region enjoys a climate more uniformly hot and moist than almost any other part of the globe, and teems with natural productions which are elsewhere unknown. The richest of fruits and the most precious of spices are here indigenous. It produces the giant flowers of the Rafflesia, the great green-winged Ornithoptera (princes among the butterfly tribes), the man-like Orang-Utan, and the gorgeous Birds of Paradise."

What had so far been collected from the archipelago and brought back to England

The archipelago has had a turbulent geological history, as evidenced by high mountain ranges, volcanoes, and earthquakes. (Lindsay Hebberd / Woodfin Camp & Associates)

was spotty. Wallace came to know the British Museum's insect and reptile catalogs almost by heart, well enough to understand how deficient the Southeast Asian collections were. No naturalists' field manuals existed for the region, and only a handful of books would be useful to him. He bought some, read others, and took copious notes.

Beyond specimens that he could study at the British Museum and the East India Museum in London, elsewhere in Britain there was a scattering of living botanical and zoological Malay curiosities. Well before Wallace's time as a young botanizer, Glasgow and Edinburgh had pitcher plants, native to Java, Borneo, and Sumatra. These plants made up for nutritional deficiencies in the soil with bright colors and smells that attracted insects, which fell into the pitcher, to be dissolved in a bath of acidic liquid and absorbed as food. By the end of the 1830s, the Zoological Society gardens in London had the Argus pheasant from Malacca and the fireback pheasant from Sumatra. And a naval expedition brought back from Celebes an animal called a babirusa. To British eyes it was bizarre, looking like a pig but with two sets of tusks, the upper set growing to as much as a foot long, curving in front of the eyes. This babirusa at sea was an opportunistic feeder: it ate six Malay birds and the brass off some marines' uniforms.

Strange stories of extraordinary birds and beasts excited Wallace's interest in the Malay Archipelago. This specimen, which he may well have seen at the British Museum as he gathered information, was identified as *Corvus indicus.* (Courtesy of the British Museum, London)

Early expeditions brought back exotic species such as the pitcher plant from Borneo and the babirusa from Celebes (now Sulawesi). (Frans Lanting / Minden Pictures; Alain Compost)

Then there was the big animal known as the orang-utan, or orang-utang, or orang-outang. It was supposed to be something like a man. From early in the age of exploration, travelers' tales had been coming back to Europe about humanlike creatures that lived on the other side of the earth—but it was not clear where. The stories came out of Africa and Asia both, and as late as Wallace's time there was confusion about orang-utans, chimpanzees, and gorillas.

The name *orang hutan* was Malay. It translated to "man of the forest" or "man of the jungle." The stories said the creature had emotions and no tail but was hairy all over. In one version, it had been banished for some unspecified transgression, to live in shame in the forest. Another version was that it could talk, but never would in front of humans because they would set it to work. Where might something so strange fit into the scheme of living things?

Dead specimens arrived in Europe, preserved in barrels of liquor. They were inspected and dissected. Linnaeus did his classifier's best with the evidence and initially put the orangutan in the same genus as man, naming it *Homo sylvestris orang-outang.*

Late in the eighteenth century, a live one from Borneo arrived in Holland and survived for a time in the natural history collection of the Prince of Orange. When it died it was dissected. By the 1830s there were some dozens of specimens in museums around Europe—full skeletons and skins, or bits and pieces. In England, the museum at Hull had a huge pair of feet. The story was that they had been in the family of a sultan on Borneo for more than a hundred years. By the size of those feet, whatever walked on them would have had to be taller than a man. In 1830 a young male orangutan arrived in London alive. It died after three days and was dissected. Beginning in 1837, the Zoological Society gardens at Regent's Park had three live females on show—not together but one after the other, because none of them survived for long, even in the heated giraffe house. They were all called Jenny. Charles Darwin saw the first one sitting in an armchair, contentedly eating an apple. The young Queen Victoria saw the third one drinking tea with a smart cap on her head, and was amused. All three Jennys were before Wallace's time as an interested observer of animals.

While he was away in the Amazon, another live specimen arrived in London, the best yet: a big male sent by the governor of Singapore. It was called Darby. Wallace never saw it; but he did give as one of his compelling reasons for wanting to go to the Malay Archipelago the chance to see the orangutan in its native habitat, and to think about its possible links with man.

The bird of paradise was another powerful attraction for Wallace. There was no creature on earth more beautiful. Its spectacular plumage had been traded for at least two

Rumors of the mysterious orangutan and bird of paradise were strong attractants for Wallace as he began his exploration. Photo on right shows bird of paradise skins from the Aru Islands in Maluku, similar to specimens Wallace viewed before embarking on his journey. (Frans Lanting / Minden Pictures; Jez O'Hare)

thousand years in many parts of Asia, as far north as Nepal and China, and for centuries west to Persia and Turkey, but not as far as Europe.

The earliest Westerners to see the brilliant plumes were sailors on the first circumnavigation of the globe early in the sixteenth century. Their sighting was in the eastern islands of the Malay Archipelago. They brought home two birds, but not alive—dead, skinned.

In the bird of paradise story that developed in Europe, with embellishments, no human ever saw the birds alive, because they never alighted on the earth. They flew high in the heavens, their plumes more precious than gold, their faces always turned toward the sun, the female laying her eggs on the back of the male, the young hatching in flight, living on dew. This life of celestial flight explained the curious fact that they had no feet. They were wanderers of paradise, descending to earth only in death.

By the start of the seventeenth century there were many bird of paradise skins in the cabinets of rich European collectors. Colonists in the Dutch East Indies shipped more skins and added more information. A Dutch scientist at the University of Leiden learned enough to be able to say that there was more than one species, and that the birds came from islands not far from Papua.

By the middle of the century, bird of paradise skins were arriving in Europe with feet just like other birds. It turned out that absence of feet had nothing to do with perpetual celestial flight, but with something more earthly: the way native hunters skinned the dead birds and prepared them for sale. Still, when Linnaeus, working from dead specimens, formally named two species for science, he called the larger one *Paradisaea apoda*, meaning "without feet."

Another curious fact caught Wallace's attention: through the end of the eighteenth century and well into the nineteenth, no European with an educated scientific eye—no naturalist—had ever observed a bird of paradise alive in its native habitat. The first such scientist to see one in flight with his own eyes was a Frenchman, René Lesson. That was in 1824, the year after Wallace was born. There had not been a scientific sighting since.

When Wallace focused his interest on the Malay Archipelago, one of the prizes of nature most firmly in his sights was the living bird of paradise.

Sea travel halfway around the world was expensive, and Wallace was hardly better off after his Amazon expedition than he had been before. But his name was beginning to be known in scientific circles, from what he had published. He applied for assistance to

The flying prau, a typical fishing vessel of the archipelago. Painting by J. C. Rappard, 1883. (Mary Evans Picture Library)

the president of the Royal Geographical Society, Sir Roderick Murchison, and on Murchison's say-so was given free passage on a Royal Navy brig headed in the right direction. Wallace was taking an assistant, a sixteen-year-old named Charles Allen. They were aboard and Wallace was already anticipating seasickness when the ship's orders were changed. War had broken out in the Crimea; it was assigned there. Murchison arranged accommodations on a steamer of the Peninsular & Orient line, a first-class cabin as far as Alexandria. Thence up the Nile to Cairo. Overland by horse-drawn omnibus to Suez (the Suez Canal was not yet built), along an excellent road dotted with camel skeletons. Then from Suez to Singapore on another P&O steamer, forty-five days altogether en route, arriving on April 20, 1854. And then the Malay Archipelago, for eight years.

At sea among the islands, Wallace covered about fourteen thousand miles, some by steamer, far more by schooner, native prau, and small boat. Like everyone else out on those waters, he had to wait upon the wind and sail before the storm. Ashore, he touched at port towns that had been centers of regional trade for hundreds, even thousands of years before Westerners first sailed the archipelago—Macassar on Celebes, Ternate in the Moluccas. And he was a brief sojourner in dozens of small tribal societies, among human beings as strange to him as he was to them. Many times he was the first white man they had ever seen.

He found food and shelter wherever he could, paying rent to village chiefs, sharing dirt floors with native families, or building his own bamboo huts. On islands east and west he slept under roofs with human heads stored in the rafters, among Dayaks in Borneo, Papuans in New Guinea. In between there were Sulu pirates at sea and Malays running amok on shore.

None of this alarmed Wallace. His view was that you could make any place sound dangerous. Take London. You could talk about the terrors of the city streets, runaway cabs, falling chimneys, garroters and mad bulls and mad dogs, and an inhabitant of tiger-infested Java or run-amok Macassar would wonder how people could be foolhardy enough to live in such a place. So with the islands. Wallace took his own measure of things and went about his business.

Singapore, his stepping-off point, was British-colonial. Elsewhere the archipelago was Dutch-colonial. Wallace needed, and got, the cooperation of Dutch civil servants and official residents. He needed, and got, letters of introduction to sultans and rajahs. Beyond that, island by island, he was dependent on the kindness of strangers, brown men and white—ship captains, traders, doctors, and miscellaneous residents—for advice

Harbor view of Singapore with trading vessels of the era. (Tettoni, Cassio, and Associates PTE Ltd. / PhotoBank)

about everything from the weather to where to find the best beetles, for the use of a house or the loan of a horse, for the destination and date of departure of schooners or praus sailing between islands.

Wallace traveled light, but not light enough to go it alone. He had his hammock, blankets, folding table, dishes, wash basin, his British comforts—tea, sugar, salt, butter, a bottle or two of wine—the rice and curry powder of the islands, and his tools of trade: guns, skinning knives, preserving fluids, collecting boxes. At any time he might have from one to three or four islanders with him, hired to carry, cook, and hunt. They came and went. His assistant, Charles Allen, did not last the distance. Wallace did not consider this a disabling loss. Early on, he had the great good fortune to run across a young Malay named Ali, who signed on and stuck with him the better part of his eight years and fourteen thousand miles, and was invaluable all the way.

As a white man, Wallace most of the time was on his own, off the edge of the map of Western experience. That map, physical and mental, most often amounted to only a coastal outline. Twenty miles from where a steamer docked, in a place where the Dutch had governed, at least in name, for centuries, Wallace might find himself in a village where the people had never seen a pin and did not know what paper was. Show people of one island a magnetic compass, and they were amazed. Show people of another island what an insect looked like through a hand lens of only a few magnifications, and amazement was magnified a thousandfold.

For some islanders away at the remote eastern end of the archipelago, it was impossible to believe that there could be a place called England. To them the name itself was gibberish—*Ung-lung, Ang-lang, Anger-lang, N-glung*. And Wallace, taller than almost every indigene who came across him, was six feet one inch of alien oddity. To spot this apparition standing in a streambed scooping at insects with a basin and spoon, or splitting open a rotting jackfruit to see what would come scurrying out, or keeping watch over a pile of dung to see what might light upon it—that was a hilarious spectacle. "One day when I was rambling in the forest, an old man stopped to look at me catching an insect. He stood very quiet till I had pinned and put it away in my collecting box, when he could contain himself no longer, but bent almost double, and enjoyed a hearty roar of laughter."

Wallace was a show. "A few years before I had been one of the gazers at the Zulus and the Aztecs in London. Now the tables were turned upon me, for I was to these people a new and strange variety of man, and had the honour of affording to them, in my own person, an attractive exhibition, gratis." Just the sight of him eating would draw crowds, like crowds watching the lions in the zoo at feeding time. Was he white all over? He bared a leg. In some places he was so rare and strange that he excited terror in man and beast. The extreme was on Celebes. "Wherever I went, dogs barked, children screamed, women ran away, and men stared as though I were some strange and terrible cannibal monster. Even the pack-horses on the roads and paths would start aside when I appeared and rush into the jungle; and as to those horrid, ugly brutes, the buffaloes . . . they would first stick out their necks and stare at me, and then on a nearer view break loose from their halters or tethers, and rush away helter-skelter as if a demon were after them."

Wallace's Biological Laboratory

Geology teaches us that the surface of the land and the distribution of land and water is everywhere slowly changing. It further teaches us that the forms of life which inhabit that surface have, during every period of which we possess any record, been also slowly changing.

WALLACE, *The Malay Archipelago*

Wallace's monumental contributions to the fields of biogeography and evolutionary biology owe a certain tribute to blind luck in his choice of the Malay Archipelago as his biological laboratory. It is one of the most geologically and geographically complex areas of the world. It is a place of dynamic change, as evidenced by more than four hundred volcanoes and innumerable earthquakes, which shake and shape the landscape, and shallow seas and deep straits that limit or completely isolate whole biological communities. More than 17,500 islands, varying in size from miniature continents to tiny atolls, and separated from continental landmasses sometimes by thousands of kilometers, are contained within its boundaries. Extreme altitudinal gradients in these islands encompass diverse habitats ranging from mangroves at sea level to glacial snowfields on mountain peaks 5,000 meters high. It straddles the equator for 5,000 kilometers, and warm, stable tropical temperatures and an abundance of rain have created huge expanses of rain forest, second in size only to those found in Brazil but equal or even greater in diversity of species. And here, two of the six major biological realms of the earth—those of Asia and Australia—come in close juxtaposition and sharp contrast. Indeed, nowhere else in the world do two realms approach so closely. For a biogeographer, a person who studies the geographical distributions of organisms, their habitats, and the historical and biological factors that produced them, the region is a gold mine. Could Wallace, the first modern biogeographer, have chosen a better place on earth to observe nature in all its infinite variety?

Wallace did ponder going to other places after the disastrous end of his trip to the Amazon, including the Philippines and Africa, but he settled on the Malay Archipelago "first, for economic reasons, because it was a ripe field to harvest for the rich collectors who bought his specimens; and second, because he believed 'man' may have originated there," says James Moore, a historian of science at Britain's Open University who has spent a quarter century researching and studying the lives of Darwin and Wallace. If Wallace had gone another route entirely and chosen a location in the temperate zone instead of the tropics, patterns affecting species distributions would have been obscured by glaciation events.

Dan Simberloff, a community ecologist at the University of Tennessee, contends that "there was some serendipity in the fact that Wallace was working in the Malay Archipelago, although it is not the only part of the world where he could have made observations that would have led him to the same views of biogeography and evolution. Had he worked in the West Indies, for example, and perhaps the Galápagos, as Darwin did, he might have been led along the same path." Simberloff's own research concerns the forces that cause a group of species to be found together in the same biological community and the factors that might lead to their extinction. Through extensive fieldwork on mangrove islands throughout the world, Simberloff has developed a deep appreciation for the lessons that islands can teach. "Being in the Malay Archipelago allowed Wallace to see within-species morphological differences between islands—crucial to his ideas of evolution. He also saw between-island differences in biological communities—crucial to various of his biogeographical ideas—and lots of evidence of the earth's dynamism. His recognition of the huge difference in communities of two nearby islands stemmed from and informed his insights on evolution, biogeography, and earth history."

Wallace, to be sure, made an educated choice. Even before he left, he was armed with questions concerning animal distributions—why and how species were found where they were. He was also well versed in the writings of other proto-biogeographers of the early 1800s, such as Charles Lyell, Joseph Hooker, Robert Chambers, P. L. Sclater, A. P. de Candolle, Henry Forbes, and Alexander von Humboldt. These men described groups of plant and animal species and their locations throughout the world, then drew up ecological maps defining regions based mainly on common adaptations to the conditions of the places the organisms were found. Although they established patterns of species distribution, creationist thinking led these men to believe that each species was created anew and placed in its location by a divine hand, ready-made to survive the climate, moisture, and other environmental attributes of their locale. Physical barriers like seas, mountains, and rivers were thought to keep each species in place.

Yet this concept of animal distribution could not explain the diversity of floras and faunas found in disparate parts of the world—equatorial Africa and South America, for example—that shared similar climate and conditions. Tropical rain forests on these two continents may look similar, but their species compositions are entirely different. Just consider the toucans of South America and the hornbills of Africa, which, though ecological equivalents and similar in looks and behavior, belong to completely different orders of birds.

Although Lyell and others believed that activity on the earth's crust constantly changed landscapes—volcanoes and molten rock raised the crust, which subsided or sank by an equal amount elsewhere—they also believed that the great continental landmasses were immutable, stable, and had always been in the same locations. If individual species were not pulled from thin air and distributed in their places, how did they manage to populate such far-flung continents and islands, places divided by vast oceans or enormous mountain ranges? Why would a divine creator not place the same species in places that had similar conditions?

Wallace, like Darwin, kept his mind open to explanations that did not invoke creationist ideas. He sought evidence to explain how and why different species came to be distributed where they were in light of past geological events, remaining aloof to the widespread notion that the physical environment alone could determine floras and faunas. Wallace used the few tools of his day—a limited knowledge of geology and the fossil record, and the distributional patterns of living, extant species—to group species into regions by common evolutionary descent.

Wallace believed that even though animals of isolated but similar environments might look the same, they could be completely unrelated, having evolved separately to the same end. Species within a region evolved from a common ancestor and could be more closely related to other nearby, but physically and behaviorally quite different, species than to their ecological equivalents in a distant region. Wallace made evolutionary maps of species distributions, a novel idea for which he sought supporting evidence by collecting in the tropics.

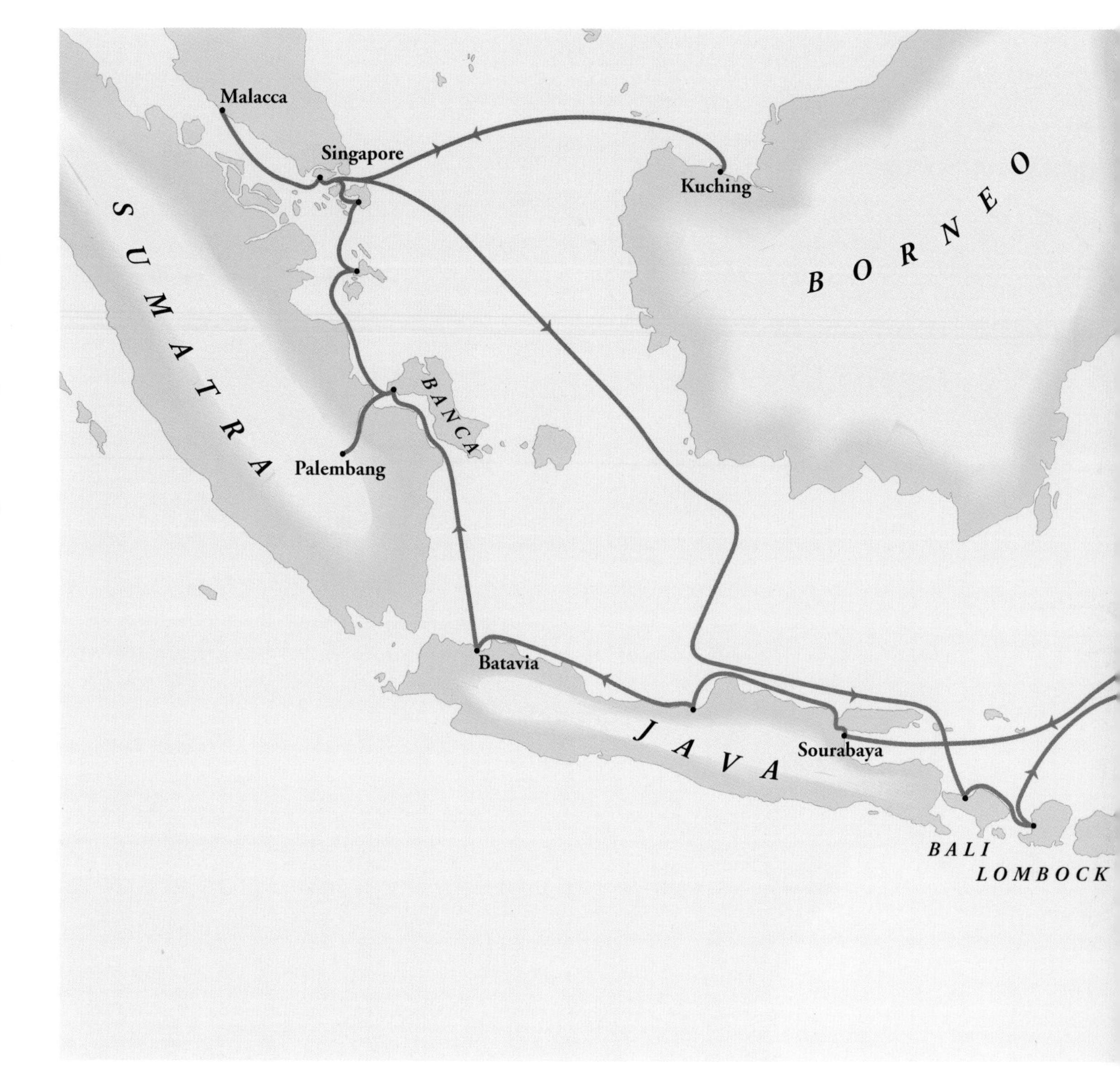

Luckily for him, the Malay Archipelago held all the answers. During his life's work, Wallace synthesized geology, geography, and study of the distribution of species and their evolution, thereby laying the foundation for modern biogeography. It would not be until almost one hundred years later, with the development of the field of plate tectonics, that his ideas were accepted and placed in their proper context: that of a history of the earth that began over 140 million years ago with the breakup of Gondwana (see p. 84).

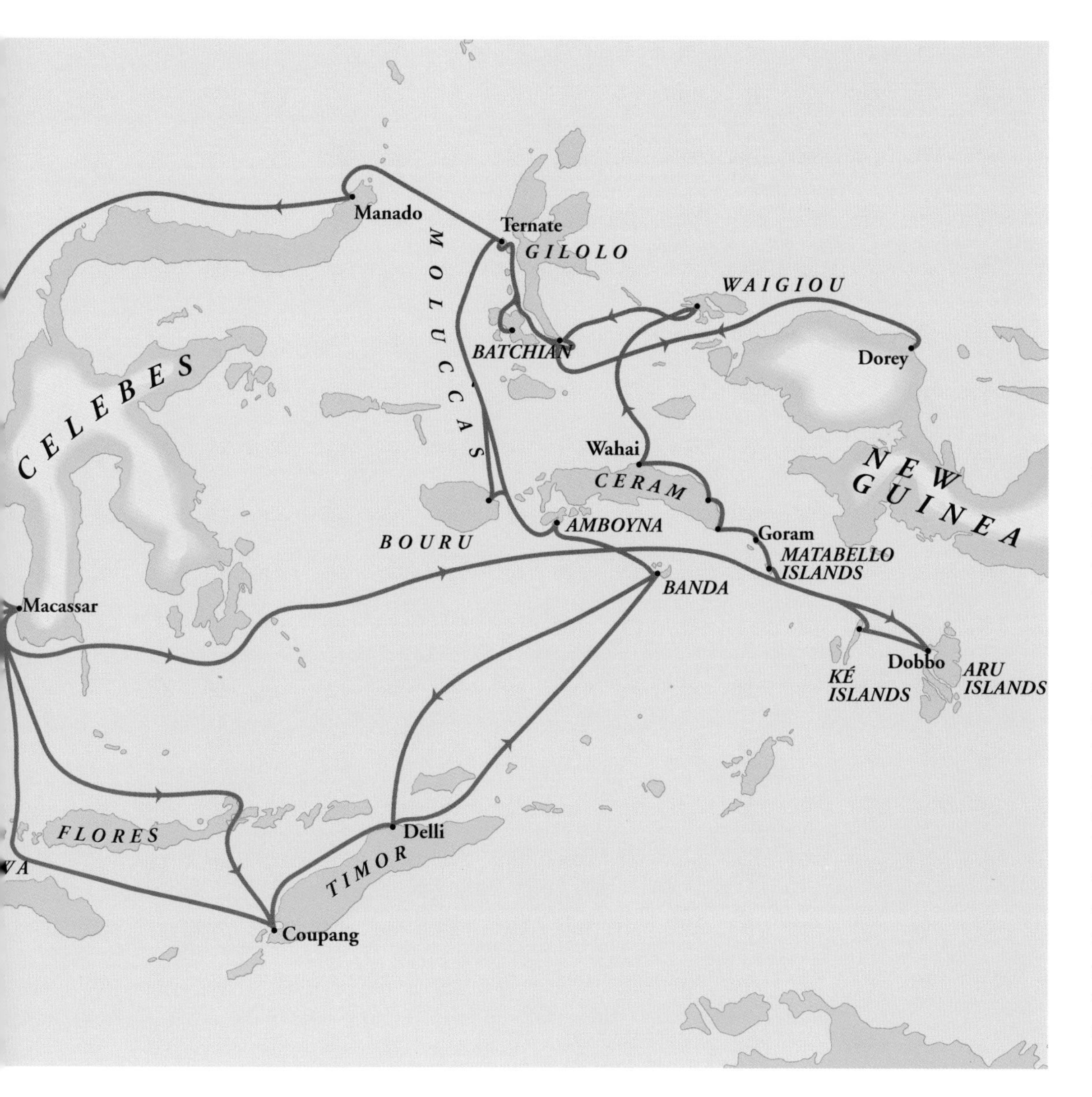

Alfred Russel Wallace traveled for eight years through the Malay Archipelago, aboard vessels ranging from mail steamers to small local fishing boats called praus. Destinations important to his collecting and thinking include, in chronological order: Singapore and Malacca (1854); northern Borneo (1855); Bali, Lombock, and Celebes (1856); Ké and Aru Islands (1857); Ternate, Amboyna, New Guinea, and Batchian (1858); Ceram, Timor, Ternate, and Gilolo (1859); Ceram, Goram, Ternate, Matabello Islands, and Waigiou (1860); Macassar, Timor, Ceram, Java, and Sumatra (1861); Singapore, before departing for London (1862). Note: all place-names are historical as used by Wallace in the description of his travels in *The Malay Archipelago*. (Modified from Wallace 1869a [1962])

The forests of Borneo are among the most species-rich habitats in the world. (Jez O'Hare)

3 BORNEO

Encountering the Orangutan

In his eight years in the islands, Wallace shifted base on average every five or six weeks. He did not start out in such rapid motion. He was at Singapore three months, in and around Malacca two months, then on Borneo a year, by far his longest stay on a single island.

Borneo was one of the biggest islands in the archipelago. It was only partially colonized by the Dutch. The northwestern region of Sarawak was controlled by an Englishman, James Brooke, known as the "White Rajah." Wallace spent many months in Brooke's area of influence.

In March 1855, as the seasonal rains were easing and the days were turning sunnier, Wallace came to Simunjon. It was the right time of year for good insect collecting, and he had found the right place. Clearings were being cut around some coal works, and it turned out that for Wallace's purposes one square mile of felled trees and decaying timber, bark, and leaves was better than a hundred square miles of virgin forest. Insects swarmed; beetles abounded. Wallace paid the Dayak natives and the Chinese mine workers one cent per piece, and in a month had more than a thousand species.

Moths were free. At a cottage of Brooke's in the mountains, Wallace sat at night out on the verandah, low-roofed and whitewashed, with a lamp and his collecting net, boxes, forceps, and pins. Some evenings only one or two moths came and he passed the time reading. On rainy nights they came by the thousands, and he caught them by the hundreds—more than 800 over just four nights; on his best night, 260; 1,386 over twenty-six nights.

He saw enormous flights of Borneo bats at dusk, black clouds of them two miles across. He sighted a remarkable butterfly, unfamiliar to him, deep velvety black, with a

Bungalow of Sir James Brooke, the "White Rajah," in Sarawak, Borneo, ca. 1858. (Mary Evans Picture Library)

collar of crimson and brilliant green spots in a curved band across the wings. It was a swift flyer, settling for but moments at puddles and muddy places. He managed to catch only two or three. They were of a new species. He named it *Ornithoptera brookeana.*

A Chinese workman brought him a frog. Its feet were webbed, and when the webs were spread they were bigger than the body. The workman told him he had watched it come down from a high tree in a slanting direction—a flying frog, the first Wallace had ever heard of.

He walked through miles of great dipterocarp forests, with trees a hundred, two hundred, three hundred feet tall. He saw the pitcher plant in many varieties growing on the mountaintops. He had drunk its liquid in Malacca, warm and full of insects but "palatable." He ate a durian, and like everyone before and after him he rhapsodized over the taste: "Its consistence and flavour are indescribable. A rich butter-like custard highly flavoured with almonds gives the best general idea of it, but intermingled with it come wafts of flavour that call to mind cream-cheese, onion-sauce, brown sherry, and other incongruities." The durian was divine to the palate, but it was not an unmitigated delight. It had a hellish smell, which has been compared by centuries of Westerners to everything from old tennis shoes to open sewers; Wallace called it "a most disgusting odour to Europeans." Another detriment was that when the fruits were ripe, they fell from high branches. They were heavy, they had spines, they could wound, even kill. Wallace stated the moral dryly: "Trees and fruits, no less than the varied productions of the animal kingdom, do not appear to be organized with exclusive reference to the use and convenience of man."

A

B

The breathtaking sights and species of Borneo's rain forest: (a) lowland dipterocarp forest; (b) fruit bats beginning their evening foraging excursions; (c) banded palm civet; (d) phosphorescent mushrooms of the forest floor; (e) treefrog in mushroom during rain; (f) luminescent insect larvae; and (g) short-horned grasshopper. (Frans Lanting / Minden Pictures; Alain Compost; Jacana / Photo Researchers, Inc.; Alain Compost; Frans Lanting / Minden Pictures; Alain Compost; Tim Laman)

C

D

E

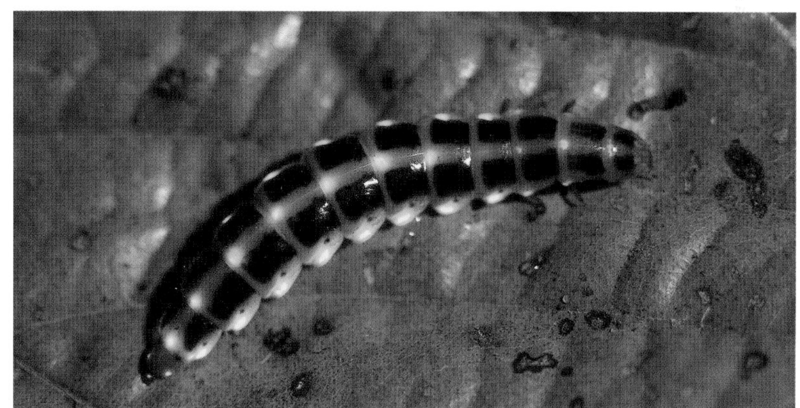

F

G

Simunjon was where Wallace sighted his first orangutan. A week after he arrived he was out collecting insects; he heard a rustling overhead, looked up, and saw a big animal covered in red hair moving from branch to branch, tree to tree, and out of sight.

Wallace intended to do more than observe. His aim was to "obtain good specimens of the different varieties and species of both sexes, and of the adult and young animals." *Specimens* meant dead orangutans; *obtaining* meant killing.

In Sarawak there was a great deal of killing of orangutans. The Dayaks hunted them using spears and blowpipes with poison darts, to keep them away from fruit-bearing trees but also for the sport of it. The Dayak phrase for hunting orangutans was the same one they used for hunting men; they had a different word for hunting monkeys. Rajah Brooke kept orangutans as pets, and he and other white men, from visiting naval officers to assistant missionaries, went out shooting them—the more the better. Brooke was interested in sending specimens back to England. So was Wallace: for science, and to sell for money to keep himself going.

Orangutans were solitary forest animals, hard to spot, and they did not like to be stalked. The males made loud noises, and females in a durian tree would break off branches of the heavy fruit and shower them down on hunters. Male or female, they were hard to kill. For Wallace with his double-barreled gun, only once was a single shot enough, and two shots rarely. Most times he had to load and reload with ball, and loose off three, four, five, eight, as many as ten blasts; and even then the animal might not be dead. It might come crashing down out of the tree cover, as one did, with both legs broken, a hip joint and the root of the spine shattered, and two bullets flattened in its neck and jaws, but still alive. Or it might refuse to fall, wedging itself in a tree fork, breaking off branches to make a kind of protective nest. If it died up there, its body would rot and be of no use to Wallace. So he would have to offer to pay Chinese to chop the tree down; or else some Dayaks would climb up to pull the carcass free, fifty or sixty feet to the first branches, hammering pegs into the trunk to support a bamboo ladder that they put together and secured as they went.

Wallace's earliest adult kills were females. The first took him five shots. The second took three to knock it dead out of a tree, and when Wallace and his Dayak of the day were getting ready to haul the body away they found its baby alive. Evidently it had been holding on to its mother when she fell. Now the mother was dead, and the baby was head over ears in the mud of a swamp the consistency of pea soup. When its mouth was cleaned out it started to cry. It clutched Wallace's beard and would not let go. Wallace carried it home.

He adopted it, and it adopted him. It was a female, weight three pounds nine ounces, fourteen inches long heel to crown, with dark brown skin, red hair, pretty little hands

Opposite left: Gliding is a unique adaptation of canopy dwellers such as the flying frog and flying lizard. (Frans Lanting / Minden Pictures)

Opposite right: A young orangutan feasts on the fruit of the durian tree. Borneo is the center of diversity for many fruits, including mango, durian, and bananas. (Alain Compost)

and feet, and a very large mouth—altogether, to Wallace's eye, "the most wonderful baby I ever saw."

It looked to be only weeks old. He gave it a box for a cradle, with a soft mat. At first it would not sleep alone, so he made a pillow out of an old sock, and the baby went to sleep hugging it. It was not yet weaned. He made it an artificial mother of buffalo skin, but when the baby tried to suck at it, all it got was mouthfuls of hair, which made it scream, and one day it almost choked. It would catch him by the trousers while he was at work and hang under his legs for a quarter of an hour at a time, trying to suck. Wallace let it suck on his finger. He rigged a corked bottle with a quill for sucking, but there was no milk at Simunjon; the Dayaks did not drink it, and neither did the Chinese or the Malays. So Wallace used thickened rice water sweetened with sugar, or coconut milk, and then spoonfuls of soaked or chewed biscuit with a little sugar and egg, and sometimes sweet potatoes. He liked to watch the baby's reactions. "It was a never-failing source of amusement to observe the curious changes of countenance by which it would express its approval or dislike of what was given to it. The poor little thing would lick its lips, draw in its cheeks, and turn up its eyes with an expression of the most supreme satisfaction when it had a mouthful particularly to its taste."

Its teeth started to come in, and it showed signs of learning to run. Wallace got it a little monkey to play with, and it liked that. He made it a ladder to hang from, to exercise and strengthen its limbs.

He was looking forward to raising it to maturity and taking it back to England with him. But it was not growing. When he had had it about three months it got sick, went into a decline, and died.

If Wallace had been a middle-class Victorian householder at home in England with a beloved pet who passed away, he might have buried it with a tombstone and an epitaph, or stuffed it and kept it around the house. He was sentimental enough in what he said about the departed: "I am sure nobody ever had such a dear little duck of a darling of a little brown hairy baby before." But at the same time as Wallace was playing with his baby, feeding it, keeping it clean, washing and wiping and rubbing it dry (which it enjoyed "amazingly"), and while he was taking daily amusement and pleasure in its curious ways and the inimitably ludicrous expressions of its little countenance, he was sighting down the barrel of his gun at other orangutans, big ones, firing shot after shot to kill them, skinning them and boiling the flesh off the bones in a big iron pot, all within sight and sound and smell of the baby.

He shot fifteen. He kept a numbered listing in his notebook, with dispositions. The baby's mother was number 7: "skeleton dry & skin in arrack." The baby's number—he never gave it a name—was 8, and Wallace's handwritten notation about its death

Young orangutan. (Jez O'Hare)

In the solitude of paradisiacal settings such as this, Wallace pondered the origin of species. (Frans Lanting / Minden Pictures)

was different from what he wrote for publication. In print, the word he used was "died." In his notebook what he wrote was: "Killed." Then: "skin in arrack with bones of limbs—scull separate cleaned." And another notation: "BM. £6"—meaning he sold it to the British Museum for six pounds.

Killing orangutans and selling the dead body parts—this was Wallace the collector paying his way. Wallace the naturalist, paying constant attention to the geography of life, was taking notes on the range and habitat of the orangutan, which was found only on Borneo and Sumatra, and only within low swampy areas of virgin forest with fruit-bearing trees.

The dimension of time as well as space was always in Wallace's thinking. All present-day animal species had forerunners. The orangutan, the chimpanzee, and the gorilla would be no exception. "With what interest must every naturalist look forward to the time when the caves and tertiary deposits of the tropics may be thoroughly examined, and the past history and earliest appearance of the great man-like apes may at length be made known."

Here Wallace was ruminating one of the great scientific questions of his century, an ultimate philosophical conundrum: the origin of species. It was never out of his thoughts.

He had found that when he was able to ponder at leisure he could arrive at a receptiveness of mind in which ideas came to him, in his words, like flashes of light. And he did his best thinking "surrounded by wild nature and uncultured man."

In the equatorial wet season, at the mouth of the Sarawak River, at the foot of the Santubong Mountains—that is, in his preferred condition of Wallacean aloneness, a single white man, more of a solitary in the tropical landscape of Borneo than an orangutan—he composed a paper. "During the evenings and wet days I had nothing to do but look over my books and ponder over the problem that was rarely absent from my thoughts. . . . Having always been interested in the geographical distribution of animals and plants, having studied Swainson and Humboldt, and having now myself a vivid impression of the fundamental differences between the Eastern and Western tropics; and having also read through such books as Bonaparte's conspectus, and several catalogs of insects and reptiles in the British Museum (which I almost knew by heart), giving a mass of facts as to the distribution of animals over the whole world, it occurred to me that these facts had never been properly utilized as indications of the way in which species had come into existence. The great work of Lyell had furnished me with the main features of the succession of species in time, and by combining the two I thought that some valuable conclusions might be reached. I accordingly put my facts and ideas on paper."

Life across space and time . . . species changing in response to changed geographical and geological conditions . . . extinctions occurring . . . new life-forms being shaped. Wallace sorted the evidence, worked his way logically through the implications, and at the end of a sequence of closely reasoned general formulations arrived at what he called a law: "Every species has come into existence coincident both in space and time with a preexisting closely allied species."

No one had advanced so far toward describing the unfolding of evolution. Wallace was still short of understanding the underlying controlling mechanism—the *how* of it—but he had come closer than anyone else.

Wallace floated his paper in the direction of England by slow boat—the only kind from Borneo—to connect at Singapore with the mails for Europe. When his insight from Sarawak was published several months later in London, in the *Annals and Magazine of Natural History*, it turned out to be ahead of cutting-edge scientific understanding. It elicited no perceptible public reaction, except that Samuel Stevens heard remarks to the effect that Wallace was "theorizing" when he should have been collecting more facts.

In biological science a tension was developing between "field" and "closet" naturalists. The field type was supposed to be content to gather specimens; the closet type was supposed to be entitled to assign meaning to what had been gathered. It never occurred to Wallace that he could not be both, do both.

His Sarawak paper did provoke one consequential private response. Charles Lyell read it, and it caused him to begin a notebook on species. It also caused Lyell to suggest to Charles Darwin that he should read it—which was actually a reaction of the greatest importance, at the highest level of scientific thought.

Evolution's Panoply

Enormous eyes and large ears are indications of the tarsier's nocturnal habits. (Frans Lanting / Minden Pictures)

Tortoise beetle peeking over the top of a leaf. Millions of species of arthropods remain to be discovered and described in Borneo. (Tim Laman)

A few hours beyond Jahi we passed the limits of cultivation, and had the beautiful virgin forest coming down to the water's edge, with its palms and creepers, its noble trees, its ferns, and epiphytes.

WALLACE, *The Malay Archipelago*

In the hearts and minds of many, the name Borneo conjures up dark and steamy images of danger in the form of impenetrable jungle, with its venomous snakes, blood-sucking leeches, and headhunters—a place where mystery abounds. For naturalists who have studied its secrets, however, it is a place of evolutionary wonder.

Borneo is the third largest island in the world, after Greenland and New Guinea. Straddling the equator, it by turns basks in sun and is deluged by rain, ideal conditions that promote the most extensive and most luxuriant stands of rain forest in all of Southeast Asia. Three mighty, sinuous rivers originate from its mountainous heartland—the Kapuas running to the west, the Barito to the south, and the Mahakam to the east—and neatly partition the island into thirds.

Political boundaries also dissect the island, a reflection of past colonial interests. In the north, the states of Sarawak and Sabah, formerly British holdings, became independent states of Malaysia in 1963, and the oil-rich independent sultanate of Brunei Darussalam lies in between. The remaining two-thirds of the island was ceded to Indonesia by the Dutch in 1949, whereupon it was divided into the four provinces of West, Central, South, and East Kalimantan. With about 9.1 million inhabitants (less than 17 people per square kilometer), the Kalimantan provinces support only 5 percent of Indonesia's population; however, they play a key

Proboscis monkeys inhabit the mangrove forests of Borneo. They have specialized guts to break down the tannin-rich leaves that form a large part of their diet. (Frans Lanting / Minden Pictures; Gerard Lacz / Animals Animals)

role in the country's economic development, with their trove of minerals and natural resources—gold, diamonds, coal, oil, and precious hardwoods.

But the real wealth, the spiritual wealth, of Borneo resides in the richness of its life, drawn from Asia and shaped by evolution. About 2 million years ago, much of the Sunda Shelf on which Borneo lies was exposed, its shallow seas retracted by Pleistocene glaciation. Land bridges connected Java, Borneo, Sumatra, Bali, and parts of the Philippines to the continent of Asia, thus facilitating the movement of plant and animal species. With the end of the glacial era and the rising of sea levels, Borneo again became an island surrounded by shallow seas.

The distinct forest types of Borneo are dictated by a combination of factors: geology, topography, elevation, and soil conditions. Mangrove forests fringe the coastline; freshwater swamp, peat swamp, and *kerangas* heath forest are found in scattered patches farther inland; and montane forests cling to the slopes of interior mountains—each harboring uniquely adapted fauna. Nowhere, though, is evolution's panoply better displayed than in the dipterocarp forests that dominate the gently rolling landscape of this island.

Lowland dipterocarp forest thrives in a climate that varies by no more than 10 degrees Celsius, with a monthly rainfall of no less than 200 millimeters. These stable conditions have given rise to more than ten thousand different species of flowering plants, one of the richest assemblages in the world. Its fauna, characteristically Asian in origin, is equally rich, and includes 222 mammal species, 420 birds, 166 snakes, over 100 amphibians, and at least 390

Stunted tree growth of montane forest in Borneo. (Michael Doolittle / Peter Arnold, Inc.)

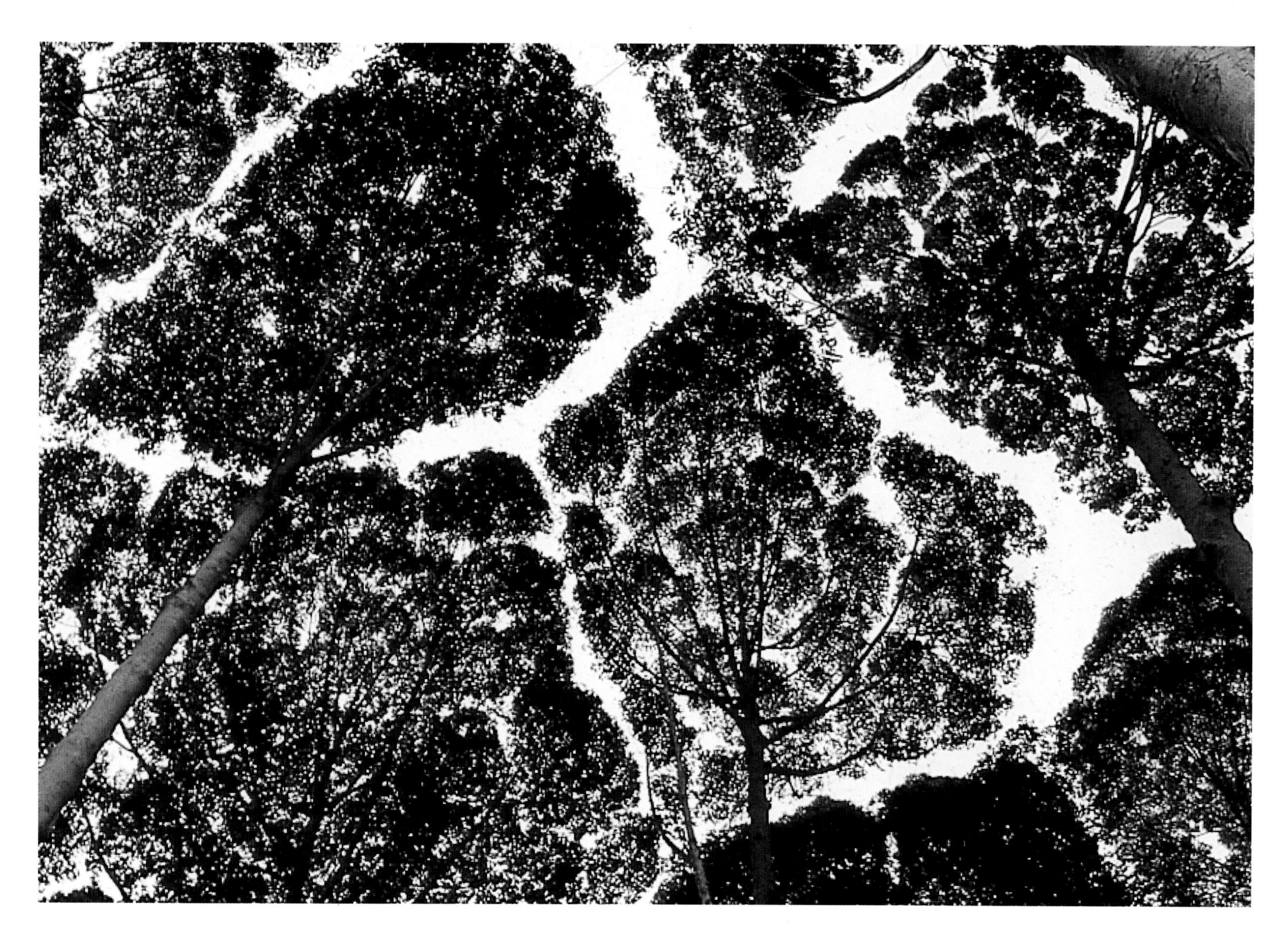

Lowland forest canopy from the ground up, illustrating the non-overlapping nature of tree crowns. (Mark Moffet / Minden Pictures)

freshwater fish species. A significant proportion of these species are unique to Borneo, a result of the evolutionary forces acting in the isolation afforded by an island domain distant from the mainland.

A first-time visitor to this forest—which, dominated by straight-boled dipterocarp trees with great anchoring buttress roots, has a canopy that towers up to 65 meters overhead—might be overwhelmed by its stature and by the complexity of forms, sounds, and smells that invade the senses: the directionless cacophony of bird calls and the incessant buzz of cicadas; the sudden flash of color as a birdwing butterfly floats into a sunspot on the forest floor; the rustle of bending branches as a gibbon leaps; the rich, sweet smell of decomposing leaf litter. Peter Ashton, an eminent Harvard botanist who has devoted forty years to understanding the species composition of dipterocarp forests, offers an astute description: "The overall feeling of walking through a dipterocarp forest is one of spaciousness—so spacious that these forests even have an echo. Quite different from the dense jungles on the poorest soils, or forests that were cut and are regenerating." Through their research, Ashton and others have given subtle order to the bewildering chaos of a habitat crowded with species.

An emergent *Koompassia* tree, bursting through the canopy. (Frans Lanting / Minden Pictures)

Trees are the dominant structure of the forest, a framework on which vertically differentiated strata rest. At the top is the main canopy with massive emergent trees, like the white-barked *Koompassia,* that burst through the layer of green. Below this are shade-tolerant lower-story trees, festooned with lianas and epiphytic ferns and orchids. And at the bottom, the forest floor, herbs and seedlings and shade-tolerant palms take hold in the few places where sun filters through the dense canopy. Each zone tends to have its own characteristic community of animals. In the tree crowns are found the frugivores and leaf eaters: raucous hornbills, orangutans, graceful leaf monkeys, and colorful barbets. The understory is the domain of the omnivores: sunbears,

A

B

Borneo harbors the richest primate fauna in the archipelago, including (a) the Bornean gibbon, (b) the red leaf monkey, and (c) the long-tailed macaque. (Roland Seitre / Peter Arnold, Inc.; Tim Laman; Tim Laman)

C

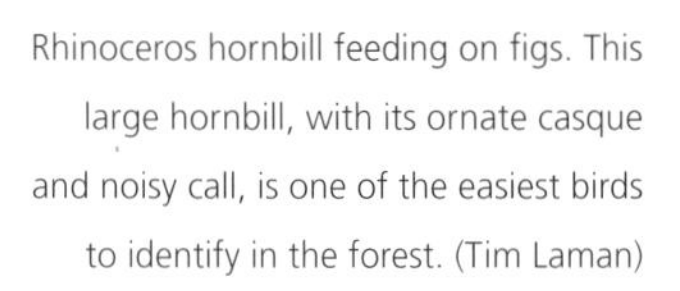

Rhinoceros hornbill feeding on figs. This large hornbill, with its ornate casque and noisy call, is one of the easiest birds to identify in the forest. (Tim Laman)

Below: Some of the smaller members of the canopy bird fauna: *(left)* a black and yellow broadbill holding a stick insect and *(right)* a green broadbill in its hanging nest. (Tim Laman)

woodpeckers, bulbuls, squirrels, and a few carnivores that range up and down tree trunks, of which the magnificent clouded leopard is one of the more awe-inspiring. Larger herbivores and omnivores—deer, elephant, and wild pigs—roam the forest floor in search of fallen fruit and tender leaves and grasses.

A temporal division also overlays the vertically stratified communities. With the setting of the sun the day shift is replaced by the night shift, composed mostly of mammal species: flying squirrels and the curious colugo, outfitted with skin flanges that enable them to glide smoothly between treetops in search of insects, seeds, and fruit; flying foxes silently fluttering among the trees looking for pungent, ripe fruit. In the understory, the tiny owl-eyed tarsier hunts large-bodied insects and small lizards and birds, and a lumbering pangolin, a scaly version of an anteater, pokes and prods for termites.

The forest's multidimensional structure is further partitioned into "niches"—finely tuned enclaves defined by the differing lifestyles and competitive interactions of species. The availability of innumerable niches in lowland rain forest permits the coexistence of even closely related species with different specializations and adaptations. For example, consider the thirty-four squirrel species in Borneo, which range in size from the giant squirrel, a 1-kilogram endemic tree-dweller, to the tiny, 20-gram pygmy squirrel. All of these species have neatly partitioned their world to reduce competition for similar foods by being either diurnal or nocturnal, by occupying different vertical strata in the forest, or by maintaining subtle differences in food preference. Similar patterns emerge within the seventeen species of fruit bat that occupy lowland rain forest, which partition their world and food sources according to body size: the largest fruit bat, *Pteropus vampyrus,* with a wingspan of about 1 meter, is capable of flying off with heavy fruit like the wild mango; medium-size short-nosed fruit bats take

Large herds of bearded pigs migrate through the lowland forests following mast fruiting of dipterocarp trees. (Frans Lanting / Minden Pictures)

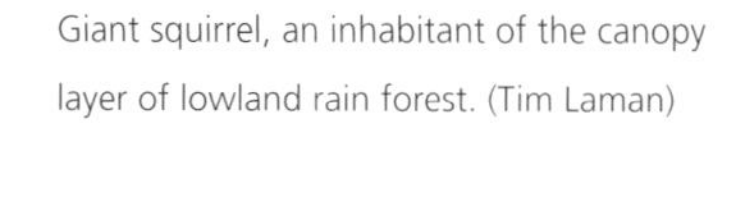

Giant squirrel, an inhabitant of the canopy layer of lowland rain forest. (Tim Laman)

A significant number of rain forest plants depend on bats for pollination, and their flowers exhibit specific adaptations to attract bats: they are usually white or dull-colored, exude copious sweet-smelling nectar, and open only at night. Many of these bat-dependent plants are also economically important. *Left:* A Horsfield's fruit bat pollinating a durian flower; *right:* a dawn bat pollinating a wild banana flower. (Merlin D. Tuttle, Bat Conservation International)

larger quantities of mid-size fruit from ubiquitous trees like *Dracontomelon* or *Syzygium;* the smallest of the fruit bats are the nectar bats, which can hover while they lap energy-rich nectar from night-blooming flowers.

This intricate web of life is held together by the interactions between plant and animal, predator and prey, parasite and host, some so tightly knit by evolution that they boggle even the most imaginative of minds. Perhaps the most elegant examples of such interactions are those that, like lock and key, have evolved among trees and their pollinators. Unlike trees in temperate forests, which are primarily wind-pollinated, most rain forest trees use animals as their sexual surrogates. Moths, butterflies, bats, and birds are all far more likely than the wind to carry pollen to the flowers of the same species of tree, individuals of which tend to be widely separated in the rain forest. This sort of pollination enhances the tree's ability to set fruit, disperse its seeds, and propagate. Wonderful examples of coevolution of flowers and animal pollinators abound: the night-blooming blossoms of the durian tree, which attract small fruit bats with their fetid fragrance and copious nectar; trumpet-shaped and brightly colored flowers that attract day-flying butterflies; beetles and flies that pollinate the rank, carrion-scented blossoms of annonaceous trees.

The most spectacularly specialized example of coevolved pollination systems is that of figs and their tiny wasp pollinators. Borneo has more than 130 species of fig trees, ranging from large strangling figs like *Ficus caulocarpa* to small, free-standing species. These figs produce prodigious amounts of fruit in one or more bursts each year, which provide an invaluable staple for almost every frugivorous

animal in the rain forest. To set this fruit, and to ensure its propagation, each species of fig depends on its own unique wasp species—minute insects no larger than the head of a pin. The cycle begins with the young fig fruit, within which are hundreds of minuscule male and female florets. As the fig develops, a few female wasps—newly hatched from nearby figs—enter a narrow opening called an ostiole at the fruit's base. Once inside, the female wasp pushes her long ovipositor down the style of a female floret and lays a single egg in the flower's ovary, avoiding longer-styled florets that her ovipositor cannot reach. In the process, the female wasp deposits pollen from the fig in which she hatched, thus accomplishing cross-pollination for her host. Wasp larvae mature with the fig fruit; within a few weeks, mature, wingless males emerge and search for female pupae to mate with before

A fig wasp emerges from the fig fruit that depends on this tiny pollinator. (Tim Laman)

Right: Although the strangler fig starts life as a tiny seed in the branches of its host tree, its roots may eventually entirely encircle the host. (Tim Laman)

Archduke butterfly feeding on fallen figs. (Tim Laman)

they, too, emerge. The male wasps then tunnel through the wall of the fig fruit to create escape tunnels for their female mates. This work also stimulates the development of male florets within the fruit, accelerating the ripening of the fig. The male wasp dies within its fig host, but female wasps, covered in fine pollen from the fig's male florets, fly off in search of another fruit of the same fig species . . . and the intricate dance begins again.

Tim Laman, an ecologist from Harvard University, has spent more than three years studying the ecology of the fig species of the lowland rain forest in Gunung Palung National Park in West Kalimantan. Hoisting himself to the lofty heights of the canopy, he explores the world-within-the-world of the fig tree. He has documented almost one hundred species of birds, mammals, and insects that count on figs—which ripen during times when other, more succulent fruits are scarce—as a diet staple. He delights in a world that few humans have seen: "The diversity of this forest is so vast that it is still relatively easy to discover new things. I discovered a new ant species quite by accident, because I found the ants stealing the fig seeds I was using in a canopy experiment!"

While the extraordinarily complex lowland rain forest is resplendent with species, the actual numbers of individuals within each species may be low; in any given hectare, there may be only a single individual of a rare tree or large vertebrate species. The entire ecosystem, so finely tuned by evolution, hangs in a very delicate balance and is thus highly susceptible to disturbance at the hand of humans. "From my botanical work in East Kalimantan," remarks Kuswata Kartawinata, a respected botanist who has worked in Indonesian forests since 1959, "I would say at least 30 percent of the plant species

Flowerpecker eating a fig. (Tim Laman)

have been lost. The main causes are human activities, in particular mechanized logging for commercially valuable dipterocarps and other species, along with conversion of forest to agricultural land and plantations. Extinctions due to natural causes over the last one hundred years or so have been insignificant."

Borneo's landscape is rapidly changing. Only 60 percent of the island remains forested, and the lowland forests, because of their accessibility to loggers and large-scale development, are today under particular threat of destruction. In the last ten years, drought and fire have ravaged vast areas; other areas have been reduced to unproductive grasslands when the thin topsoils of cleared forestland have washed away.

The indigenous Dayak people, who for centuries have lived and harvested the natural bounty of their Bornean homeland, believe that the forest tree is a metaphor for the origin of life, and that animal spirits must be invoked and assuaged for protection against evil. The Dayak's lesson is a valuable one: humanity must learn to forestall the destruction of a land that evolution has so richly endowed.

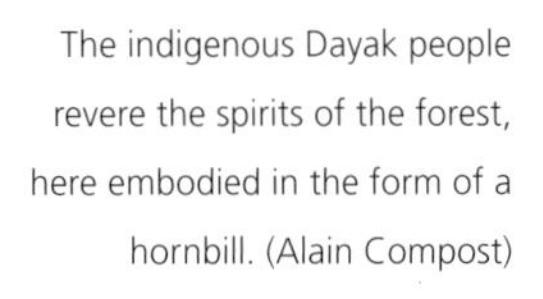

The indigenous Dayak people revere the spirits of the forest, here embodied in the form of a hornbill. (Alain Compost)

The Wild Man of the Forest

Now it seems to me probable, that a wide extent of unbroken and equally lofty virgin forest is necessary to the comfortable existence of these animals. Such forests form their open country, where they can roam in every direction with as much facility as the Indian on the prairie, or the Arab on the desert; passing from tree-top to tree-top without ever being obliged to descend upon the earth.

WALLACE, *The Malay Archipelago*

Simia, Pithecus, Papio, Pongo, satyr, ape, pygmie: the orangutan has been referred to by all of these names in the course of over three hundred years of study. Despite its many monikers, the animal still remains one of the most misunderstood and mysterious of the great apes. The intrigue of finding the missing link between animals and humans fueled stories of the "wild man of the forest" that began to circulate in Europe in the 1500s when the first Western travelers returned from the Malay Archipelago. Native legends added to the allure: Western explorers were told that an *orang hutan,* or "forest dweller," was formerly a human who had been banished to the forest and, though still capable of speech, refused to talk for fear of being put to work.

Confusion about the scientific name and classification of the orangutan ensued when European traders began sending back specimens of chimpanzees and gorillas in addition to orangutans, listing their country of origin as Angola—the main trading port in Africa—regardless of where the specimens originated. As a consequence, the name Pongo, derived from a native Congolese name for the gorilla, *mpongwe,* came to be associated with all Southeast Asian and African apes. Linnaeus unknowingly con-

Peat swamp forest in Tanjung Puting National Park in Central Kalimantan, one of the forest habitats in which orangutans are found. During the wet season, water floods into the forest to a depth of 0.5 to 1.5 meters. Acidic humus buildup stains the water black, and peat with 65 percent organic content can be 20 meters deep. (Ecopix / PhotoBank)

Orangutan female with infant. Young can be dependent on their mothers for up to seven years. (Frans Lanting / Minden Pictures)

Opposite: Adult male orangutan displaying large cheekpads and throat sac—an imposing and formidable countenance. (Kevin Schafer / NHPA)

tributed to the confusion by formally classifying the orangutan in 1766 as *Simia satyrus,* and the chimpanzee as *Simia satyrus indicus,* considering it a black-colored subspecies of the orangutan.

Not until the late 1770s, when the first live orangutans were captured from Sumatra and Borneo and described in detail by Dutch anatomists and naturalists, did its true origins become known. The official scientific taxonomic identity of the orangutan was finally established in 1929, when it was formally classified as a distinct genus and species of the family Pongidae, which includes gibbons, chimpanzees, and gorillas. It is now recognized as *Pongo pygmaeus:* the great red ape of Southeast Asia. The species was further divided into two subspecies, based on subtle physical and genetic differences: *P. pygmaeus abelii,* from Sumatra; and *P. pygmaeus pygmaeus,* from Borneo. A larger, prehistoric race, *P. pygmaeus palaeosumatrensis,* was identified as well, discovered from fossils which revealed that orangutan ancestors were once common throughout Java and roamed areas as distant as southern China. The fossil evidence also indicated that the orangutan was an important food source for Pleistocene humans, a fact that probably contributed to the extirpation of this species from all areas except the relatively uninhabited rain forests of north Sumatra and Borneo.

Wallace was one of the first to contribute information about the orangutan in the wild, describing in detail its rain forest habitat, the way it moved through the forest, its feeding and nesting behavior, and its reaction to humans and other potential threats. He dutifully recorded these traits even as he killed and collected fifteen specimens to be shipped back to London for anatomical study.

By the 1890s, interest in the orangutan had waned. The animal was regarded as lazy and slovenly, largely because of the lethargic behavior of captive orangutans. Its evolutionary stature in the great chain of being was devalued as anthropocentric research shifted to the more socially complex and lively chimpanzee, now considered the closest evolutionary relative to humans. Even as the orangutan slipped into relative obscurity, it continued to be captured and removed from the wild; some two hundred animals were exported for display in zoos and private bestiaries around the world between the 1890s and the 1960s.

In the 1960s interest in the orangutan was rekindled, primarily because of reports that the species was on the verge of extinction. Although the orangutan had been legally protected since 1925 by the Dutch and then the Indonesian government (it was the first animal in Southeast Asia to be granted formal protection), commercial logging was destroying vast tracts of its rain forest habitat, and illegal trade had burgeoned. Cursory surveys speculated that as few as four thousand orangutans remained on the face of the earth. Spurred by this news and motivated by conservation ethics, ecologists and primatologists launched extensive surveys and field studies. The details of the orangutan's idyllic lifestyle in the wild began to be revealed.

Rising at dawn from a nest of folded leaves and supple branches, the orangutan begins its daily routine. It tidily relieves itself over the side of its tree nest, then makes its way through the forest canopy at a leisurely pace, using its long arms and handlike feet to move between branches. Up to 95 percent of its typical daytime activity consists of traveling, feeding, and resting in midday siestas. The rest of the time may be spent with the occasional orangutan it happens to encounter, and then finding a tree to make a new nest to bed down in after sunset. "Primate fieldwork has been glamorized by news articles and movies of famous primatologists," says John Mitani, an anthropologist at the University of Michigan. "But does anyone really know what it's like to do observations on an orangutan that can go to sleep in the middle of the day for over six hours, 40 meters up in a tree?" Mitani's studies

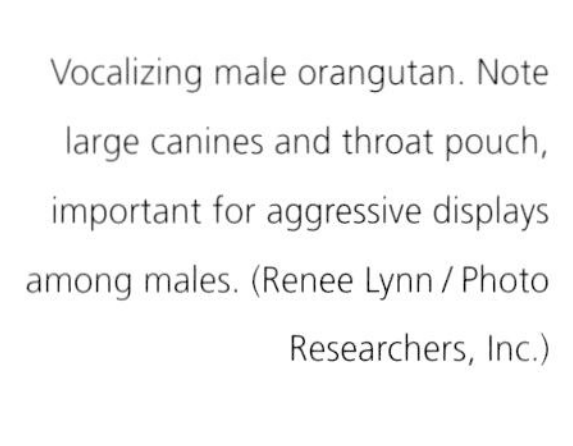

Vocalizing male orangutan. Note large canines and throat pouch, important for aggressive displays among males. (Renee Lynn / Photo Researchers, Inc.)

of communication among the great apes have provided new insights into the different vocalizations typical of each species. He found that long calls—the deep, throaty baritone bellow of adult male orangutans, which can be heard up to 800 meters away—transmit a wealth of information, such as identifying the caller and warning away competitor males.

Orangutans, unlike other great apes, are solitary by nature; a mother and her infant, which can remain dependent for as many as five to seven years, typically constitute the largest social unit. The rare encounter with another orangutan may result in greetings, play, or aggressive interactions. When two big males meet and they perceive each other as competitors for mates, they are likely to react with spectacular intimidation displays: branches are shaken, broken, and thrown, and each male rises up on his haunches to exhibit his great lumbering bulk. A "kiss squeak"—a sharp kissing sound that ostensibly serves to frighten away the competitor—generally launches such energetic displays. If that fails, the fighting can escalate to grappling and biting, a struggle sometimes fought to the death.

The orangutan is arboreal and rarely descends to the ground from the great heights of the rain forest. It feeds on leaves, young shoots, flowers, insects, wood pith and bark, and, rarely, small vertebrates and bird eggs. But the single most important component of its diet is fruit from hundreds of different tree species widely dispersed in the forest. This food source has greatly influenced the orangutan's social system. Cheryl Knott, an anthropologist at Harvard University, explains: "The most common fruit trees preferred by orangutans in Asian forests tend to be smaller and more widely dispersed than those fed on by chimpanzees in African forests. These smaller trees cannot support large groups of animals, which compels the orangutan to forage alone rather than in large groups like chimps."

Knott and her team of assistants have logged more than twelve thousand hours of observation time over three years studying the orangutan population of Gunung Palung National Park in West Kalimantan, Indonesian Borneo. "The great apes provide a window into our human origins. By knowing how we are similar to other great apes and how we have diverged, we can begin to understand what led to the changes during human evolution," remarks Knott. Her research is focused on how orangutan reproduction and behavior have adapted to the rain forest environment; she combines "old-fashioned fieldwork, chasing after orangutans much like Wallace did, with new scientific techniques." When following orangutans, Knott rises before dawn and patiently waits below an orangutan nest, collecting samples of urine below the nest for later laboratory analysis of reproductive hormones. She then shadows the individual all day, collecting samples of discarded fruits and other foodstuffs and taking detailed notes on every movement of her target animal. Although orangutan movement through the canopy seems slow and deliberate, tracking individuals from the ground can be difficult. "Sometimes we have to scramble over slippery rocks and up muddy slopes, thrash our way through rattan thickets, and trudge through peat swamps. They can travel over 1 kilometer in a day before they make their night nest." Like other researchers, Knott discovered that orangutans process an incredible amount of phenological information on each of the trees in their home range, constantly assessing flowering and fruiting stages, and sometimes travel in a determined, straight line to a particular tree that they anticipate will be in full fruit.

Knott has measured the effects of seasonal changes in fruit availability on orangutan reproduction, physiology, and behavior. "When fruit is abundant—particularly during a mast fruiting, when all of the dipterocarp trees are producing nutrient-rich seeds and when many other high-calorie fruits are available—orangutans can take in ten times more calories than in a period of low fruit availability. I've found that the female's nutritional status affects hormonal levels, thus probably influencing her ability to reproduce." Female orangutans become sexually mature at around twelve to fifteen years of age, and births are usually spaced at eight-year intervals. With such a low birthrate over an estimated life span of forty to fifty years, the growth of orangutan populations is limited by female fecundity. The female's ability to reproduce ultimately depends on the quality of the forest habitat where she lives. "Understanding the orangutan in its environment is essential for determining whether viable populations can be maintained in shrinking reserves and increasingly degraded habitats," states Knott.

Although careful recent surveys reveal that perhaps twenty thousand orangutans still inhabit the forests of northern Sumatra and Borneo, the species remains on the brink of extinction. The threats to this animal, and especially to its forest home, have not subsided. Perhaps as much as 90 percent of the orangutan's original habitat has been destroyed, and rain forests continue to be logged at an unprecedented rate. Massive forest fires in Borneo and Sumatra have taken their toll as well; during the devastating fires in 1997, for example, many orangutans died, and hundreds more were driven from the forest to villages in search of food, where they were either killed or taken captive as pets. Black market trade of the red ape continues, despite international treaties that ban such commerce. Programs to rehabilitate captive orangutans and reintroduce them into the wild have had very limited conservation value, primarily because of the paucity of relatively intact, unpopulated habitats and because introducing captive animals can threaten wild ones with new diseases. Knott and others think that the best way to ensure the orangutan's survival in the wild is to protect its forest habitat. Some, however, wonder whether it is not already too late for the great red ape.

Mount Agung presides ov
the island of Bali. (Jez O'K

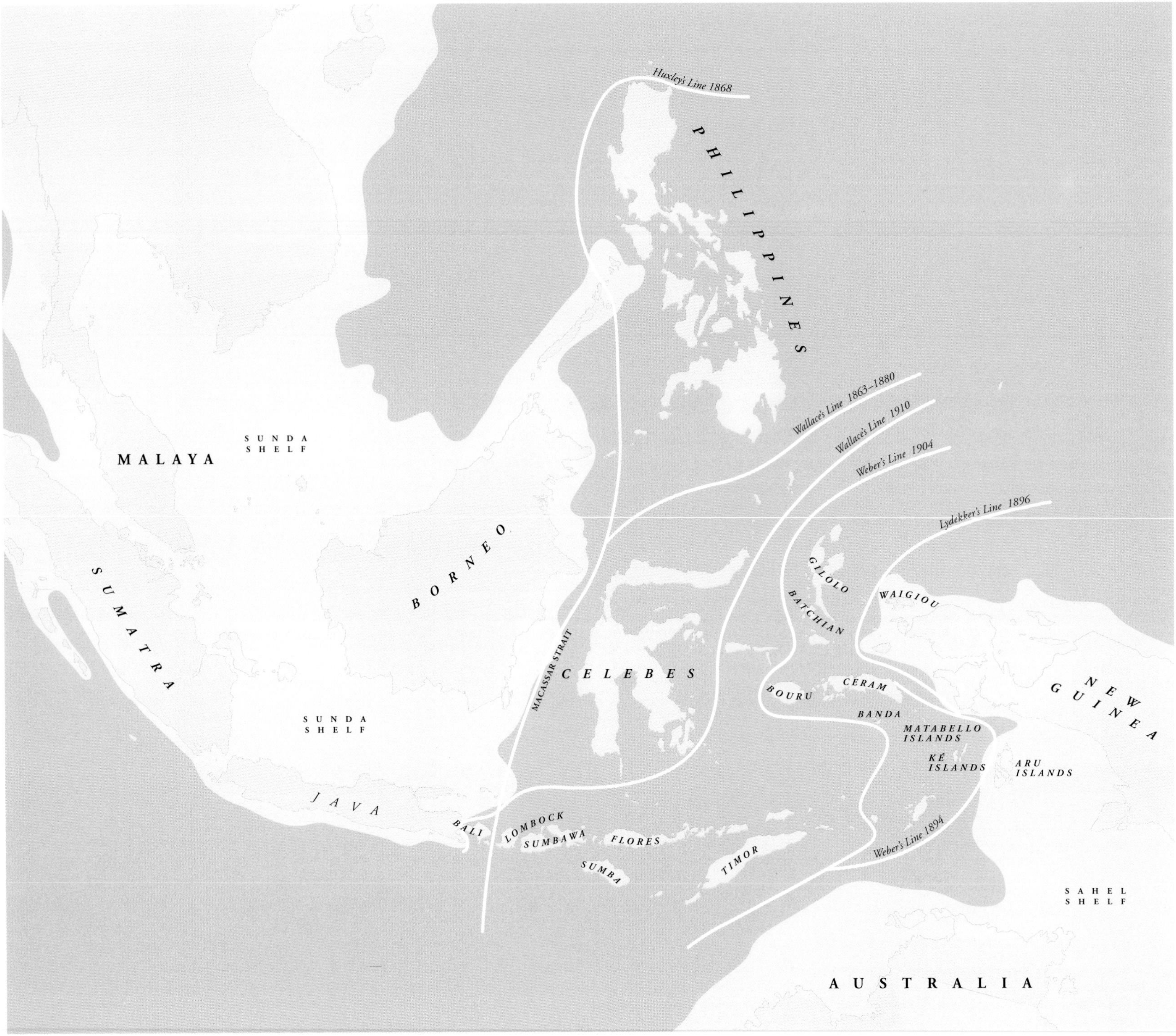

Huxley's Line 1868
PHILIPPINES
Wallace's Line 1863–1880
Wallace's Line 1910
Weber's Line 1904
Lydekker's Line 1896
MALAYA
SUNDA SHELF
SUMATRA
BORNEO
MACASSAR STRAIT
CELEBES
GILOLO
BATCHIAN
WAIGIOU
NEW GUINEA
BOURU
CERAM
BANDA
MATABELLO ISLANDS
KÉ ISLANDS
ARU ISLANDS
SUNDA SHELF
JAVA
BALI
LOMBOCK
SUMBAWA
FLORES
SUMBA
TIMOR
Weber's Line 1894
SAHEL SHELF
AUSTRALIA

BALI, LOMBOCK, AND CELEBES 4

Wallace Draws the Line

"The relations of animals to space and time, or, in other words, their geographical and geological distribution and its causes"—this was one of Wallace's great preoccupations. He intended to follow it up systematically: "I have set myself to work out this problem in the Indo-Australian Archipelago and I must visit and explore the largest number of islands possible and collect materials from the greatest number of localities, in order to arrive at any definite results."

The number of possible islands was in the thousands. And distances between islands could be very great. As Wallace observed, voyages in the archipelago were "commonly reckoned by weeks and months." Yet it happened that one of Wallace's shortest voyages—and one that was more or less accidental—produced another far-reaching scientific formulation.

After Borneo, Wallace intended to sail from Singapore to Celebes. He could not get direct passage. He had to go via Bali and Lombock.

Singapore to Bali was twenty days at sea. Wallace made the passage on a schooner owned by a Chinese merchant, with a Japanese crew and an English captain. He had only a couple of days ashore on Bali. The part of the island where he spent his time was beautiful, and he was amazed to see a sophisticated irrigation system for growing rice; but intensive cultivation everywhere meant that he could not expect to pick up much in the way of natural history. He did some more or less offhand bird shooting and butterfly hunting, and then left for a pleasant sail across a narrow channel to Lombock.

There he saw boys catching dragonflies with bird lime on a stick, to fry and eat. He came upon skeletal remains of natives—murdered or executed—in little bamboo enclosures, with clothes, pillow and mat, and betel nut box. The word was out that a ra-

Opposite: Wallace's famous line, first drawn in 1863 through the deep but narrow Macassar Strait between Borneo and Celebes, separates the two major biological realms of Asia and Australia. Subsequent lines were drawn, reflecting debate surrounding Wallace's ideas, but Wallace's original line has withstood the test of time. Islands in the shallow seas of the Sunda Shelf were once connected to mainland Asia, whereas those on the Sahel Shelf were connected to Australia. Those islands in the deeper seas between the shelves were isolated for longer periods of time. (Modified from George 1964)

The sophisticated irrigation system used for growing rice in Bali exemplifies an enduring relationship between the people and the land. (Jez O'Hare)

jah was going to be taking heads for temple offerings to guarantee a good rice crop. There was a false alarm about a man running amok. Wallace was told that there were no ghosts on the island, but that men could turn themselves into crocodiles to eat their enemies. Death by crocodile was the punishment for adultery. A man and a woman were to be tied back to back and thrown into the sea to be eaten. Wallace went for a long walk in the country until it was all over.

Out collecting, Wallace saw and shot at fruit pigeons, doves, kingfishers, bee-eaters, flowerpeckers, ground thrushes, black cuckoos, king crows, golden orioles, honeysuckers, cockatoos. The strangest birds were members of the family Megapodiidae. They were about the size of small hens, with big feet and long curved claws. They hatched their eggs not by sitting on them but by building incubating mounds of dirt, stones, sticks, leaves, and rotting wood. Dozens of birds would make a common mound, as much as six feet high and twelve feet across, with all the eggs inside.

Megapodes were singular creatures. But it struck Wallace that there was a greater strangeness about the Lombock birds taken as a group: they were vastly different from the birds of Bali.

How could this be? The distance between the two islands was short, a bird flight of only fifteen or twenty miles. Yet on the evidence of Wallace's own eyes, there was a greater

A

B

C

D

(a) The Bali starling and (b) the gold-whiskered barbet of Borneo, representative of Asiatic bird species, are found west of Wallace's Line, while (c) the black palm cockatoo of Irian Jaya and (d) the rainbow bee-eater of Sulawesi are examples of the Australian species east of the line. (Alain Compost; Tim Laman; Alain Compost; Klaus Uhlenhut / Animals Animals)

View of Wallace's Line looking west from Mount Rinjani, Lombok, to Mount Agung, Bali. The small Gili Islands, shrouded in clouds, lie in the narrow strait between the two islands. (Jez O'Hare)

difference between bird populations on either side of that narrow channel than would be seen in voyages of weeks or months elsewhere in the world.

No barbets east of Bali; no honeysuckers west of Lombock. This was a shorthand way of saying that the Bali bird population looked like the bird population of Java, meaning that its connections were ultimately with the Asian continent. By contrast, the connections of the Lombock bird population were ultimately with the Australian continent.

For Wallace this was an entirely new way of seeing, and it turned into an entirely new way of thinking. Later, after he had sailed the length of the archipelago, observing and ruminating, he wrote to his friend Henry Bates: "There are two distinct faunas rigidly circumscribed, which differ as much as those of South America and Africa, and more than those of Europe and North America. Yet there is nothing on the map or on the face of the islands to mark their limits."

Wallace was the one who drew the dividing line. And for the rest of his scientific life he kept reworking the implications of this insight. Others did too. In the twentieth century, Wallace's Line has been studied and debated as much as any one subject in biological science. It is the boldest single mark ever inscribed on the biogeographical map of the world.

Between Lombock and Celebes Wallace was only three days on the water. This short voyage landed him much deeper in strangeness. Over a five-year period he was in and out of Celebes four times, and when after eight years he left the archipelago for good, he was still not sure that he understood the island. For decades afterward he kept shuttling back and forth in his mind, trying to decide where Celebes belonged in his arrangement of zoogeographic regions—which side of his boundary line it should properly be on.

Borneo and Celebes were close to each other, without any great intervening physical barrier or climatic difference. Yet Celebes had parrots and Borneo did not, which put Borneo on the Asian side of Wallace's line, Celebes on the Australian. Upon further reflection, though, Wallace decided that Celebes was the easternmost remnant of a huge ancient Indo-Asian continent. But after further thought he changed his mind again: the hundred-fathom line of ocean depth ran between Borneo and Celebes, and Celebes was to the east of it; that put it in the ancient Australian region, which seemed to be confirmed by evidence from animal, bird, and insect life. However: Celebes pigeons had affinities with both east and west. Yet again: even though Celebes was a big island, the size of Java, and was in the central part of the archipelago, and thus might logically be

Coral reefs, mangrove estuaries, and riverine forests, all ecosystems that thrive along Sulawesi's convoluted coastline, are but a few of the many habitats found on this unique island. (Rio Helmi / Auscape)

imagined to have a great number of bird species, it did not. And of the bird species that it did have, a great number were not found elsewhere. Wallace considered the island of Celebes "one of the most interesting in the world to the philosophical ornithologist."

So it was to the philosophical zoologist as well. Animal groupings found on islands on both sides of Celebes were not found on Celebes itself. Where Celebes animals did show affinities, it was often not with close islands but with places far away—New Guinea, Australia, India, Africa.

Celebes was a single large island with only a few smaller ones nearby. Perhaps it had never been connected to any ancient continent. Perhaps it preexisted other islands. All by itself it could be considered a major division of the archipelago. The word for it was "anomalous."

Celebes was poor in numbers of species but rich in peculiar forms, and for Wallace on the ground it was a place of unusual experiences. On this equatorial island he climbed high mountains to altitudes where the morning temperature was in the low sixties. He saw fantastically sculpted limestone caves and stalactites, a stupendous waterfall, volcanic springs of boiling mud, and sulphurous blowholes. An earthquake rocked his house and almost made him seasick. He came down with malaria; his companion Ali did too. He paid some children to bring him insects, and other children shot birds for him with blowpipes and clay pellets. He collected rare *Ornithoptera* and swallowtail butterflies. He did well with beetles. He acquired babirusa skulls. And he went hunting maleo birds.

The maleo that Wallace was after, the Celebes brush turkey, was a kind of megapode. It came as much as fifteen miles to a black volcanic sand beach to dig holes and lay eggs as big as teacups. Celebes villagers came as much as fifty miles to take the eggs. Wallace came twenty miles to shoot and skin the birds. For five days he ate the flesh of the ones he killed, and he found their eggs delicious. He accumulated twenty-six skins, then left with his two helpers and a guide, on foot, cutting a path for miles in the forest through ferocious tangles of rattan.

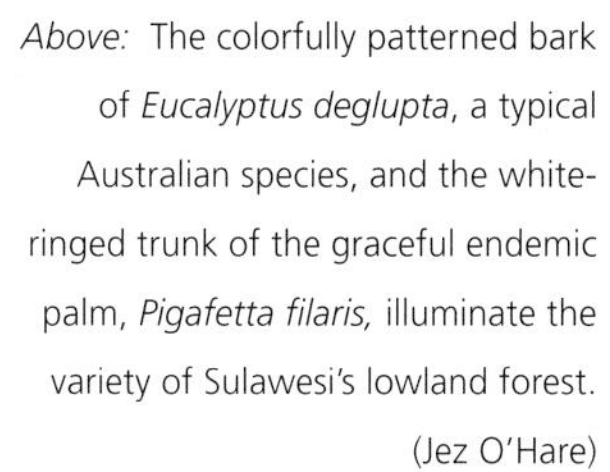

Above: The colorfully patterned bark of *Eucalyptus deglupta*, a typical Australian species, and the white-ringed trunk of the graceful endemic palm, *Pigafetta filaris,* illuminate the variety of Sulawesi's lowland forest. (Jez O'Hare)

Above right: The vivid eclectus parrot, a resident of Sulawesi's forests, attests to the Australian flavor of the bird fauna found here. Raucously screeching flocks flying overhead were once a common sight. (F. Stuart Westmoreland / Photo Researchers, Inc.)

Right: The comical stripe-faced bat, an unusual endemic chiropteran species of Sulawesi. Endemism is rare among larger bat and bird species capable of long-distance flight and dispersal. (Alain Compost)

Wallace's Vision of Continents

The 6-kilometer-wide Lombok Strait, where two major biogeographic realms approach each other most closely in the archipelago. This breathtaking view looks east across Wallace's Line from Bali to Lombok. (Jez O'Hare)

From many points of view these islands form one compact geographical whole, and as such they have always been treated by travellers and men of science; but a more careful and detailed study of them under various aspects, reveals the unexpected fact that they are divisible into two portions nearly equal in extent, which differ in their natural products, and really form parts of two of the primary divisions on earth.

WALLACE, *The Malay Archipelago*

A remarkable pattern emerged from Wallace's ponderings of the thousands of mammal and bird specimens that he collected during his travels throughout the archipelago. In two regions, he found two distinct faunas separated by neither major climatic differences nor major physical barriers—only a narrow band of water. The faunas of the great islands of Sumatra, Java, Borneo, and Bali appeared to have been derived from Asian stock. He surmised that these islands, all lying on the shallow Sunda Shelf, where water depth never exceeds 50 meters, were once connected to the Asian continent, perhaps during glacial periods when such shallow seas virtually vanished, leaving land bridges among the islands that permitted free diffusion of species. A similar shallow-water area, the Sahel Shelf, connected the islands of New Guinea and Australia. This concordance could explain the similarities among fauna in those two places. The islands between these shelves—Sulawesi, Maluku, and a smattering of smaller ones—lay in deep ocean; their faunas, moreover, seemed more associated with that of Australia.

In 1863, Wallace drew a line separating these faunas: it ran through the Makassar Strait between Borneo and Sulawesi and the narrow strait separating Bali and Lombok. It seemed almost inconceivable that the two major biological realms of Australia and Asia could be separated by a water barrier so slim—28 kilometers at its widest point and 6 kilometers at its narrowest. Yet to the west of this line could be found rhinoceros and elephants, to the east tree kangaroos and phalangers: species representing two distinct mammalian groups, placentals and marsupials, which diverged as early as 120 million years ago. Although the difference among birds is less evolutionarily dramatic, the line also appeared to separate typically Asian families like pheasants and hornbills from the more typically Australian cassowaries and birds of paradise.

Wallace's bold idea touched off a firestorm of debate among supporters and detractors that lasted well into the 1940s and resulted in many new lines named after their conceivers. The distinguished British biologist T. H. Huxley (who named the line after Wallace in 1868) analyzed the distribution of birds and placed the Philippines to the west of the line. In 1896 R. Lydekker repositioned the line such that it traced the western edge of the Sahel Shelf, thus delimiting a western boundary of a strictly Australian fauna, unlike Wallace's Line, which designated the eastern edge of the Asian fauna. In 1904 M. Weber constructed a line between Wallace's and Lydekker's, representing a divide where the ratio of Asian to Australian mammals and mollusks was 50:50. In 1945 the evolutionist Ernst Mayr contended that Wallace's Line separated a continental fauna rich in numbers of species from the relatively impoverished island faunas of Sulawesi and Maluku. Mayr concluded that if any line should be drawn to differentiate the Asian and Australian realms, the most sensible would be Weber's, which traced a boundary of faunal balance.

Each of these investigators, including Wallace himself, based his arguments primarily on the distributions of birds and mammals, the more obvious and extravagant creatures of nature's design. Had plants, freshwater fish, reptiles, and most insect species been subjected to the same scrutiny, no clear patterns between west and east, Asian and Australian realms, would have emerged. Had Wallace been a botanist, a significant chapter of the history of science would have had to be rewritten.

One point everyone agreed on, including Wallace: the island of Sulawesi was an anomaly. Although more of its bird and mammal species appeared to have been derived from Asian than from Australian stock, endemic species (that is, species found nowhere else on earth) prevailed. Going by the evidence, Wallace surmised that Sulawesi had been isolated far longer than any of the Sunda Shelf islands, giving evolution a much greater opportunity to shape a unique fauna. Wallace himself had trouble deciding on which side of his line Sulawesi should go; in 1910 he revised his 1863 line and placed Sulawesi to the west. The faunas of the smaller Maluku islands, though more strongly Australian in flavor, were also rich in endemic species. With too many lines and little consensus, the anomalous islands of Sulawesi and Maluku in the deep seas between the Sunda and Sahel Shelves were relegated to "Wallacea"—a zone of transition, an evolutionary limbo.

It was not until the late 1970s, when the field of plate tectonics emerged, that the clamor and controversy finally died down. From this new science we learned that the earth's crust is composed of rigid plates, akin to a giant jigsaw puzzle, except that the plates are capable of shifting position relative to one another. Plates are formed at mid-ocean ridges in a process called seafloor spreading; they are destroyed in a process called subduction, when one plate slides beneath another. Where a continent attempts to subduct, mountains are created. One such subduction region is Southeast Asia.

About 180 million years ago, the earth's large landmasses, Gondwana (present-day South America, Africa, Antarctica, India, Australia, and New Guinea) and Laurasia (present-day North America, Europe, and Asia) began to split apart from one

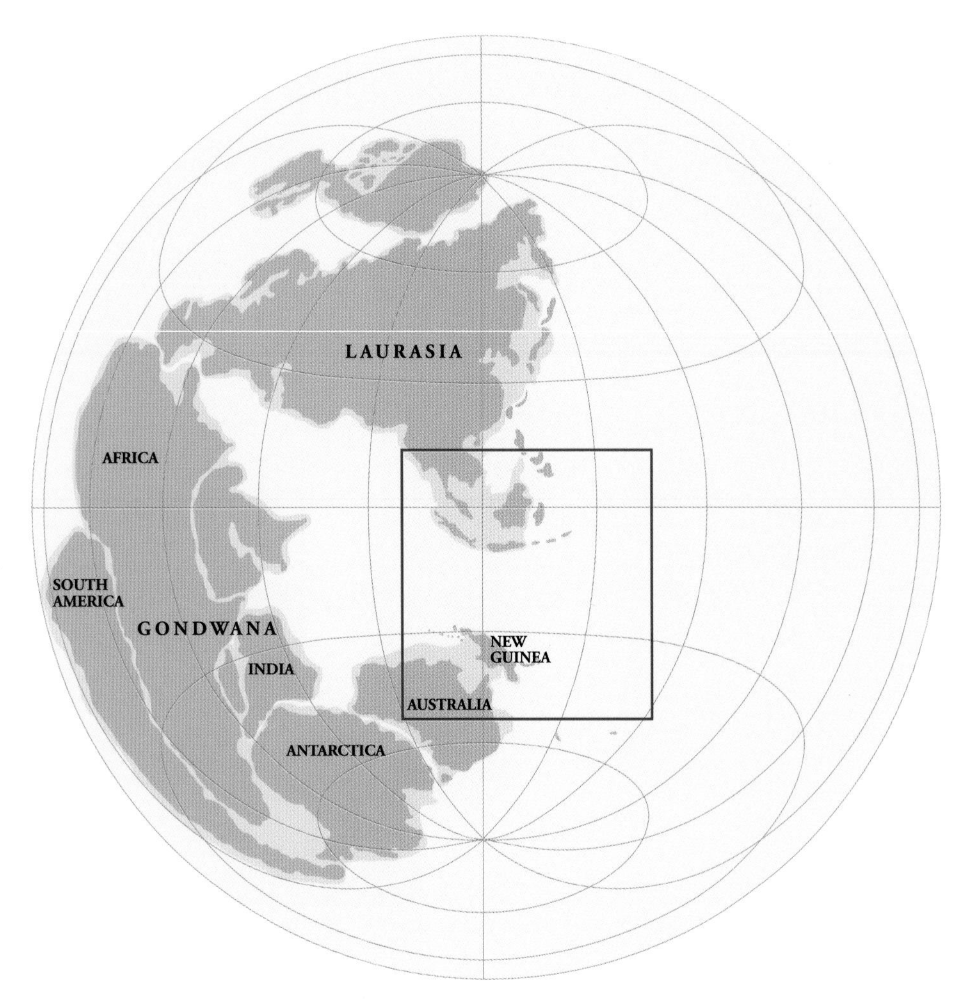

Global reconstruction showing the world at 180 million years ago, referencing the present region of Southeast Asia and Australia (shown in detail and over time on facing page). During this time Pangaea, consisting of the large landmasses of Gondwana (present-day South America, Africa, Antarctica, India, Australia, and New Guinea) and Laurasia (present-day North America, Europe, and Asia), began to break up. (Modified from Audley-Charles et al. 1981)

another. Laurasia drifted to the northern hemisphere, while most of Gondwana remained in the southern. With respect to the formation of Indonesia, several important events occurred since then. Gondwana began breaking up around 140 million years ago; a huge chunk, India, drifted northward and eventually collided with the Asian portion of Laurasia. At this time, the great Indonesian islands of Sumatra and Borneo were still attached to the Southeast Asian portion of Laurasia. About 53 million years ago another piece of Gondwana, comprising Australia and southern New Guinea, broke away and drifted northward toward Southeast Asia. This left behind the last vestige of Gondwana—now Antarctica. During the past 15 million years, the Gondwanan Australia–New Guinea chunk collided with the Laurasian portion of Southeast Asia in the area of Wallace's Line. This relatively recent collision brought two originally separate floras and faunas exceedingly close together; it also created the island of Sulawesi. The enigmatic character of Sulawesi could now be explained: it truly is a geological and biological anomaly—a strange chimera of the two ancient continents of Laurasia and Gondwana.

At last the mechanics of Wallace's Line could be explained. Vindication? With the new evidence from plate tectonics has come acceptance among biogeographers of Wallace's prescience in remarking that "vast changes in the surface of the earth" have occurred. The line that bears his name does indeed separate two ancient biological realms, regardless of its exact placement. Wallace was the first to invoke evolution and the earth's geologic history to explain a major natural phenomenon, at a time when no one else had even ventured to presume that continents could move. Wallace's place in the history of science becomes more secure with every new refinement to our understanding of the natural world.

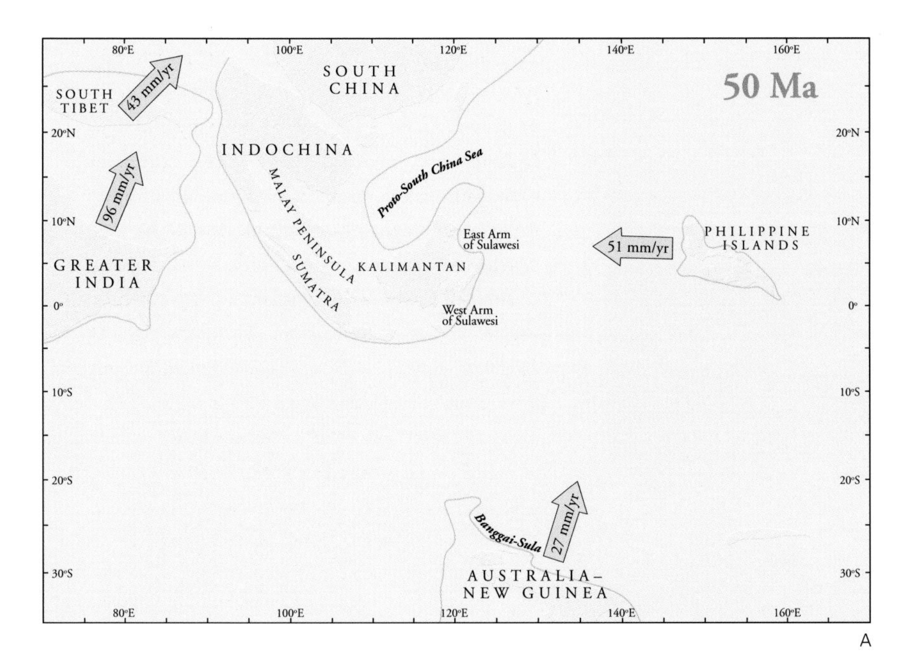

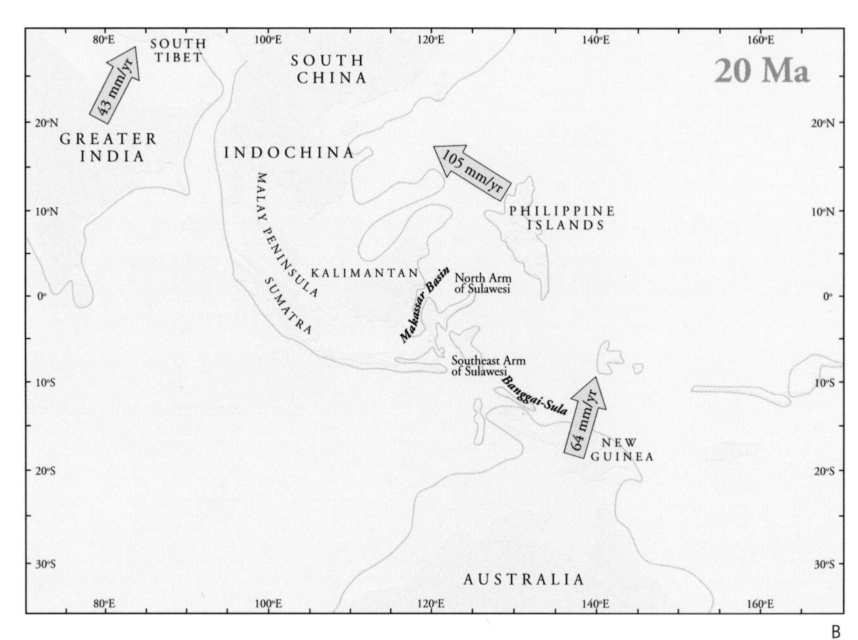

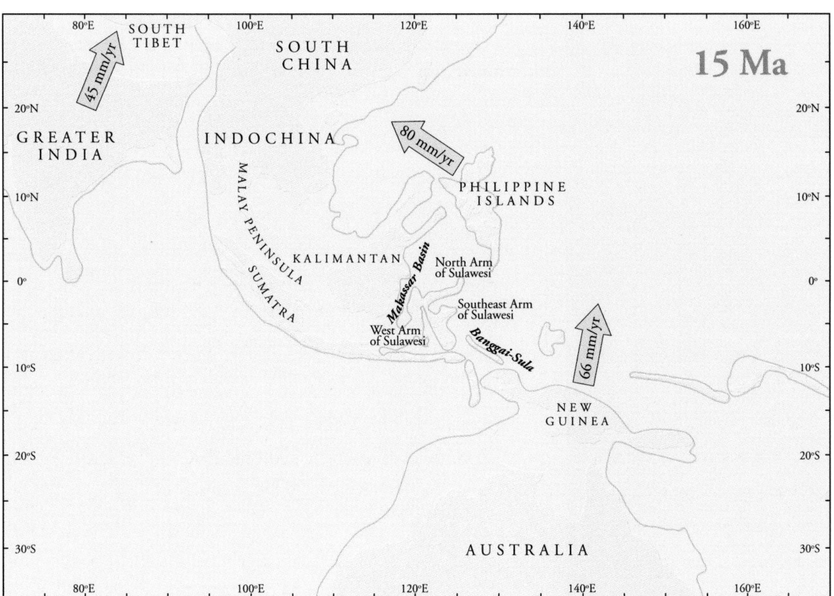

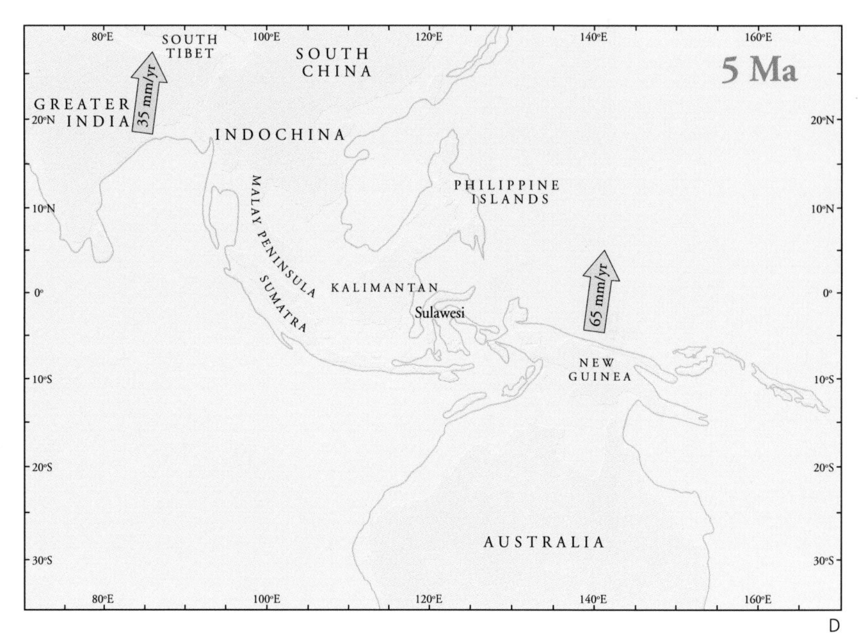

Tectonic plate movements leading to the formation of Southeast Asia: (a) Approximately 50 million years ago (Ma), a piece of Gondwana, comprising Australia and New Guinea, begins a rapid drift northward toward Southeast Asia. (b) By 20 Ma, the Makassar Strait separating Borneo and Sulawesi has widened; the northern arm of Sulawesi rotates clockwise, and the southeastern arm drifts toward it. (c) About 15 Ma, the Gondwanan Australia–New Guinea chunk collides with the Laurasian portion of Southeast Asia in the area of Wallace's Line; the island of Sulawesi is formed when western, northern, and southeastern arms are joined. (d) At 5 Ma, Indonesia begins to resemble its present-day form. (Modified from Lee and Lawver 1995; with permission from Tung-Yi Lee and Lawrence Lawver and the Plates Project at the University of Texas at Austin)

Forest still blankets the mountainous backbone of Sulawesi. (Jez O'Hare)

Sulawesi: Island Enigma

Situated in the very midst of an Archipelago, and closely hemmed in on every side by islands teeming with varied forms of life, its productions have yet a surprising amount of individuality. While it is poor in the actual number of its species, it is yet wonderfully rich in peculiar forms, many of which are singular or beautiful, and are in some cases absolutely unique upon the globe.

WALLACE, *The Malay Archipelago*

Flying over the island of Sulawesi on a cloudless day, a modern traveler will marvel at the island's strange shape and rugged topography. The 227,000-square-kilometer island, slightly smaller than England and Scotland combined, is the fifth largest in the archipelago. It is composed of four mountainous, bizarrely shaped peninsulas that splay out in separate directions like a body with arms and legs akimbo, so that no point on the island is farther than 90 kilometers from the sea. This shape gives Sulawesi the longest coastline relative to land area of all the islands of the archipelago.

The island's odd shape, as well as its anomalous natural history, is rooted in its geological history. About 160 million years ago, western Sulawesi, comprising what are now the southwestern and northern peninsulas, was part of land that lay submerged beneath the great Tethys Sea, between the vast continents of Laurasia and Gondwana. About 90 million years ago, most of eastern Sulawesi was part of land beneath either the Indian Ocean or the Philippine Sea. Between 100 and 40 million years ago, western Sulawesi, occupying a position adjacent to Borneo, was experiencing a surge of volcanic activity, and toward the end of this period began to split away from Borneo, forming the oceanic seaway now called the Makassar Strait. Much later, about 15 to 10 million years ago, a sliver of crust called the Sula Spur broke off of northern Australia and began moving northward, pushing up ocean floor ahead of it and eventually colliding into the longer western portion of Sulawesi like a spearhead. A jagged mountainous backbone, the result of violent rifting and uplifting, marks where the two parts crashed together. (See page 85.)

Sulawesi, or Celebes as it was called until the early 1900s, remained an enigma to Wallace because of the unique flora and fauna he encountered there. And indeed, the island is the ultimate biological laboratory, one in which many agents of change played a role: geological and climatic events that allowed flora and fauna of the Australian and Asian biological realms to mix; the rare chance arrival and survival of a wayfaring seed or animal colonist; the forces of natural selection acting on species in isolation. The number of endemic species attests to the strength of these forces: if species capable of long-distance dispersal are factored out (like some bats and birds that can cross large bodies of water), an astonishing 98 percent of mammal and 27 percent of bird species are endemic—found nowhere else in the world except Sulawesi. This remarkable array of species is due mainly to Sulawesi's wide variety of available habitats—a variety rich with opportunities for adaptation and diversification, such that new species form much more frequently than on other oceanic islands, which are generally habitat-poor.

The island's surrounding seas boast an assortment of marine ecosystems, including some of the best seagrass beds and coral reefs in Indonesia and the Taka Bone Rate atoll, third largest in the world. Great expanses of mangrove forest still fringe much of the coastline. There are thirteen freshwater lakes scattered throughout the island, some more than 5 square kilometers in surface area and over 500 meters deep; their crystalline ecosystems harbor unique fish species and support a host of migratory birds.

Although these largely unexplored aquatic ecosystems may hold many evolutionary secrets, it is the denizens of Sulawesi's forest habitats that best exemplify natural selection's eccentricities. At least fourteen types of forest are found on Sulawesi, each reflecting different soil types, rainfall patterns, and altitudinal zones. In the rainshadow of the mountains in the central peninsula, where annual rainfall, at 300 millimeters, is the lowest in Indonesia, monsoon forest can be found. Little is known about this kind of forest, which can withstand long dry seasons, and today only a few remnant patches remain, the rest replaced by grasslands for grazing cattle.

On the eastern and southeastern peninsulas, unusual forests grow on limestone and ultrabasic soils, generally the most nutrient-poor of all soils. Characterized by twisted trees of short stature, these forests are low in their total number of plant species but rich in endemic ones that have adapted to the harsh conditions. Pitcher plants abound, for they have escaped a dependency on soil nutrients by remarkable adaptations that enable them to capture and extract nutrients from animal prey. They hang gracefully from branches, sprouting from every nook and cranny. It is eerily quiet here; only a limited number of birds and mammals can survive on the few fruit produced by starved plants.

Yet these forests are alive with another life-form. In the strange limestone outcroppings of Bantimurung in southeastern Sulawesi and in the rugged

The strange limestone outcroppings of Bantimurung in South Sulawesi rise abruptly from near sea level. (Jez O'Hare)

Hordes of butterflies (*Graphium sarpedon*) often congregate around mineral-rich puddles. (Alain Compost)

ultrabasic forests of Morowali in Central Sulawesi, butterflies drift like petals from the forest canopy, everywhere brightening the green backdrop with their brilliant colors. Wallace was fascinated by the butterfly fauna of Bantimurung and the many endemic species he found. He was particularly taken by a spectacularly large swallowtail, *Graphium androcles:* "As this beautiful creature flies, the long white tails flicker like streamers, and when settled on the beach it carries them raised upwards, as if to preserve them from injury."

"Local butterfly catchers call this species *kupu-kupu rudal,* or 'the missile,' due to the speed of its flight," comments Duncan Neville, a Nature Conservancy biologist, who, like Wallace, delights in the many unusual forms of Sulawesi's butterflies.

He is working with villagers in Central Sulawesi to develop a butterfly farming cooperative, which will provide a dependable income source while protecting the butterflies' forest habitat. The cooperative captures adult butterflies from the forest and breeds them to produce live pupae for export to butterfly houses around the world. "The market is growing, particularly in the U.S., and Sulawesi is an attractive source for this market because of its large number of endemic species, like the spectacular iridescent blue *Papilio blumei* that we are now focusing on," explains Neville.

A fifth of Sulawesi lies above 1,000 meters, cloaked in montane forest dominated by oaks, chestnuts, conifers, and rhododendrons. At the highest levels, the forests become stunted and elfin. Tree trunks are covered in verdant, thick layers of moss and epiphytes, products of the perpetual mist that enshrouds the forest. One of the relatively few mammal species that has adapted to the colder and more austere conditions of Sulawesi's higher altitudes is the mysterious Sulawesi civet, one of the least-known and perhaps rarest carnivores in the world. Chris Wemmer, director of the Smithsonian Institution's Conservation and Research Center, embarked on a mission in the 1970s to find and study the civet in the wild. Unlike other civets, which are primarily arboreal fruit-eaters, the Sulawesi civet has a "peculiar set of characteristics, including curiously etched or pitted canines, a short stout head, and a tooth battery suggesting a predilection for meat," notes Wemmer. In three years, Wemmer and his colleagues collected only circumstantial evidence of this species: feces laden with rodent bones, and a photograph of a civet caught in a camera trap (a device that captures an animal's image when a trigger is tripped in the path of the camera). Wemmer and others fear that the Sulawesi civet, like several other Sulawesi endemics, may become extinct before its habits have become known to science.

The anoa, an endemic wild cattle species, feeds on fallen fruit and tender leaves and shoots on the forest floor. (Alain Compost)

The vast majority of Sulawesi's biological wealth is contained in lowland rain forest below an elevation of 1,000 meters. Most of the animals found in these forests are derived from primitive Asian stock, having made their way to the island from the western part of the archipelago in rare colonization events. By rafting, swimming, or crossing transient land bridges linked to Asia, Sulawesi's animal ancestors beat incredible odds to reach its shores, succeeding perhaps only once every 500,000 years. According to the fossil evidence, even a prehistoric pygmy elephant and a giant tortoise once existed on Sulawesi, having survived a long-distance swim from other islands, only later to succumb to forces of natural selection, and perhaps human hunting pressure, that led to their extinction.

Although most of Sulawesi's endemic mammals are rodents, some of the larger forest residents are among the most remarkably odd in the entire archipelago. The babirusa, or "pig-deer," resembles a small, smooth-skinned hippo, except for the enormous, wickedly recurved canine tusks of the male that protrude upward through its snout. The anoa, a diminutive but fiercely aggressive distant relative of the Asian water buffalo, quietly roams the forest floor looking for young leaves, fruit, grasses, and ferns. One of the world's smallest primates, the 10-centimeter-long, 100-gram spectral tarsier is a particularly endearing representative of Sulawesi's oddities. Three species are known to exist on the island.

The best example of how speciation events can rapidly occur on an isolated island like Sulawesi is its endemic monkey, the crested black macaque. At least seven morphologically distinct taxa (taxonomists are not yet sure if all of these are separate species) are distinguished, based on subtle differences in coat color and size of the Mohawk-like crest of stiff hairs that adorns the monkeys' heads, and by their occupation of distinct and separate areas. Their closest living relative is the pig-tailed macaque, common throughout Southeast Asia.

Nora Bynum, currently academic director with the Organization for Tropical Studies, spent more than three years researching the ecology and systematics of the Sulawesi macaques. As she puts it, they are "a natural experiment in evolution within the genus *Macaca.* They exhibit astounding morphological variation but are very similar in terms of behavior and ecology. The million-dollar question is, what accounts for all this differentiation? Did the Sulawesi macaques descend from a single common ancestor or from more than one, in the short 600,000 years since colonization from western Indonesia? Did geographic barriers isolate groups? To make this intricate puzzle even more complicated, we know that hybridization between groups has occurred." She adds: "Unfortunately, a full exploration of the fascinating evolutionary history of the Sulawesi macaque may be an unrealizable goal in a world that is changing more rapidly than we can fathom."

An extraordinary mammal reveals Sulawesi's ancient ties with Australia: the cuscus, primitive marsupials not found on islands west of Sulawesi. The two species, the dark brown bear cuscus and the dwarf cuscus, are tree-dwelling nocturnal animals that creep slowly through the forest canopy in search of leaves, fruit, and insects, wrapping their long, prehensile tails around branches for support.

Top: The crested black macaque, one of at least seven endemic macaques found in Sulawesi. (Tui de Roy / Minden Pictures)

Bottom: The endemic bear cuscus, a primitive marsupial, reveals Sulawesi's ancient ties with Australia. (Tui de Roy / Roving Tortoise)

A female red-knobbed hornbill takes a final look before sealing herself into a hollow tree nest cavity for four months to brood her young. (Tui de Roy / Roving Tortoise)

Right: A male red-knobbed hornbill delivers fruit to his mate inside the nest cavity. This is the larger of the two endemic hornbills in Sulawesi. (Tui de Roy / Roving Tortoise)

Sulawesi is a bird-watcher's paradise. Lowland forests resound with the cacophonous calls of birds with exotic names that hardly begin to describe them: cerulean cuckoo shrike, purple-bearded bee-eater, black sunbird, ivory-backed woodswallow, blue-backed parrot, golden-mantled racquet tail, ornate lorikeet, hanging parrot, spot-tailed goshawk, crimson honeyeater, yellow-billed malkoha, purple-winged roller, fiery-browed mynah. Perhaps the most spectacular of these endemics is the red-knobbed hornbill, which boasts a prominent casque, or horny extension of the bill: red in the male and yellow in the female. These massive yellow bills with brilliant red chevrons at the base are accentuated by the electric-blue skin of eye rings and gular throat pouch.

The epitome of peculiarity in both behavior and appearance is the maleo, a chicken-sized bird with

A

B

C

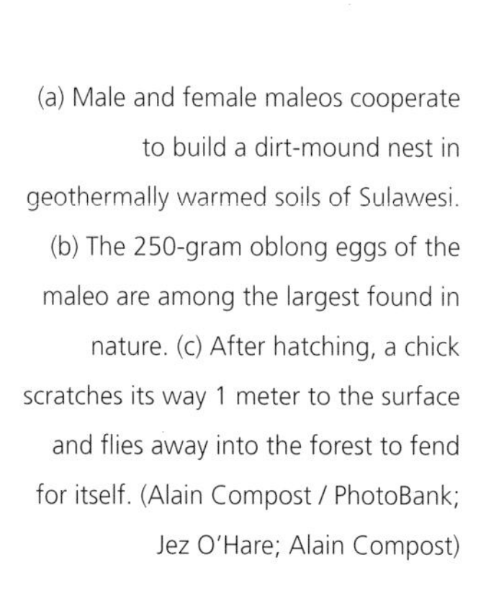

(a) Male and female maleos cooperate to build a dirt-mound nest in geothermally warmed soils of Sulawesi. (b) The 250-gram oblong eggs of the maleo are among the largest found in nature. (c) After hatching, a chick scratches its way 1 meter to the surface and flies away into the forest to fend for itself. (Alain Compost / PhotoBank; Jez O'Hare; Alain Compost)

Fishtail and woka palms thrive in small sun gaps in the forest. (Tui de Roy / Minden Pictures)

powerful legs and feet. The maleo is adorned with black and salmon plumage, a stout tail, and a grotesque black helmet. Mated pairs roam traditional nesting sites in a stately walk, looking for a suitable spot to dig a hole and deposit a single, gigantic egg. These communal nesting sites are often used for decades; they are found in soils heated by geothermal springs or in exposed beach sand warmed by the sun. Eggs are laid up to a meter deep and covered with soil by the frantic scratchings of both parents. After a long incubation period of two to three months, the chick hatches, struggles for days to reach the surface, and miraculously bursts out of its soil entombment. Once free, the chick flies off into the forest to begin life completely on its own.

Wallace's Line, which runs through the deep Makassar Strait that separates Borneo and Sulawesi, is an effective barrier not only for animals but also for plants. Only 7 species of dipterocarp trees are found here, compared to 267 in Borneo. Yet the forest is distinctive for its other denizens as well: majestic *Eucalyptus* trees from the Australian realm; *Diospyros celebica,* the blackest and hardest of all ebonies; an abundance of beautiful palms of myriad

Lontar palms flourish in forest openings.
(Alain Compost)

forms, like the fishtail *Caryota* or the fast-growing, slender *Pigafetta filaris* palm that bursts through the lowland forest canopy in search of sunlight.

Rattans, climbing palms that either twist on the ground then climb straight up tree trunks or dangle like ropes across forest clearings, are ubiquitous in Sulawesi's forests. Their delicate bright green fronds provide an inviting contrast to the relentless deep greens of the forest, but they can be treacherous. "Walking in a forest dominated by rattan demands your constant attention," says Steve Siebert, an ecologist from the University of Montana. "Rattan stems, shoots, and leaves are armed with dense whorls of long, stiff spines, some recurved like grappling hooks for climbing, all of which can shred clothing or skin in seconds." Siebert has studied Indonesia's rattan species for ten years; he lives by two rules: "First, never grab anything without looking, even when falling. Second, if ensnared by rattan, stop and back up."

For the past few years, Siebert has been conducting research on the ecology and management of economically important rattan species in Central Sulawesi, a major source of cane for Indonesia's multimillion-dollar furniture industry. He has worked closely with local villages to collect, identify, and determine the abundance and distributions of rattan species and their role in village economies, where rattan is used for dozens of purposes including furniture, baskets, tools, and dyes, as well as cash income. Siebert is exploring ways to cultivate and harvest rattan from hillside farms and parklands, thereby creating economic incentives to conserve forest habitat.

Humans have long been an integral part of Sulawesi's natural tapestry. Great kingdoms of agriculturalists and seafaring peoples have flourished here, harvesting the wealth of Sulawesi's lands and seas. The island is rich in artifacts of the vanquished: mysterious stone megaliths; intricate wrought iron urns, drums, and statues; delicate porcelains from Asia, a staple of ancient trade. Its complex cultural evolution is today reflected in the faces of its people and the diversity of its languages and dialects.

Things are changing very fast in Sulawesi. Until recently, human impact was limited by rugged terrain: agriculture stopped at the foot of the mountains. With rapid population growth, however, including the arrival of immigrants from more crowded islands in the archipelago, more and more mountainsides are being stripped of their forest cloaks. Only ragged patches of lowland rain forest remain. Aboriginal peoples, like the Wana of Central Sulawesi, who know the forest and its rhythms, are, like so much of the island's fauna, finding refuge in isolated parklands set aside by modern laws. In less than two hundred years, the enigma that so fascinated Wallace has been plundered, and many of Sulawesi's evolutionary secrets have been lost without ever having been told.

Tidore Island in the Moluccas.

(Kal Muller / PhotoBank)

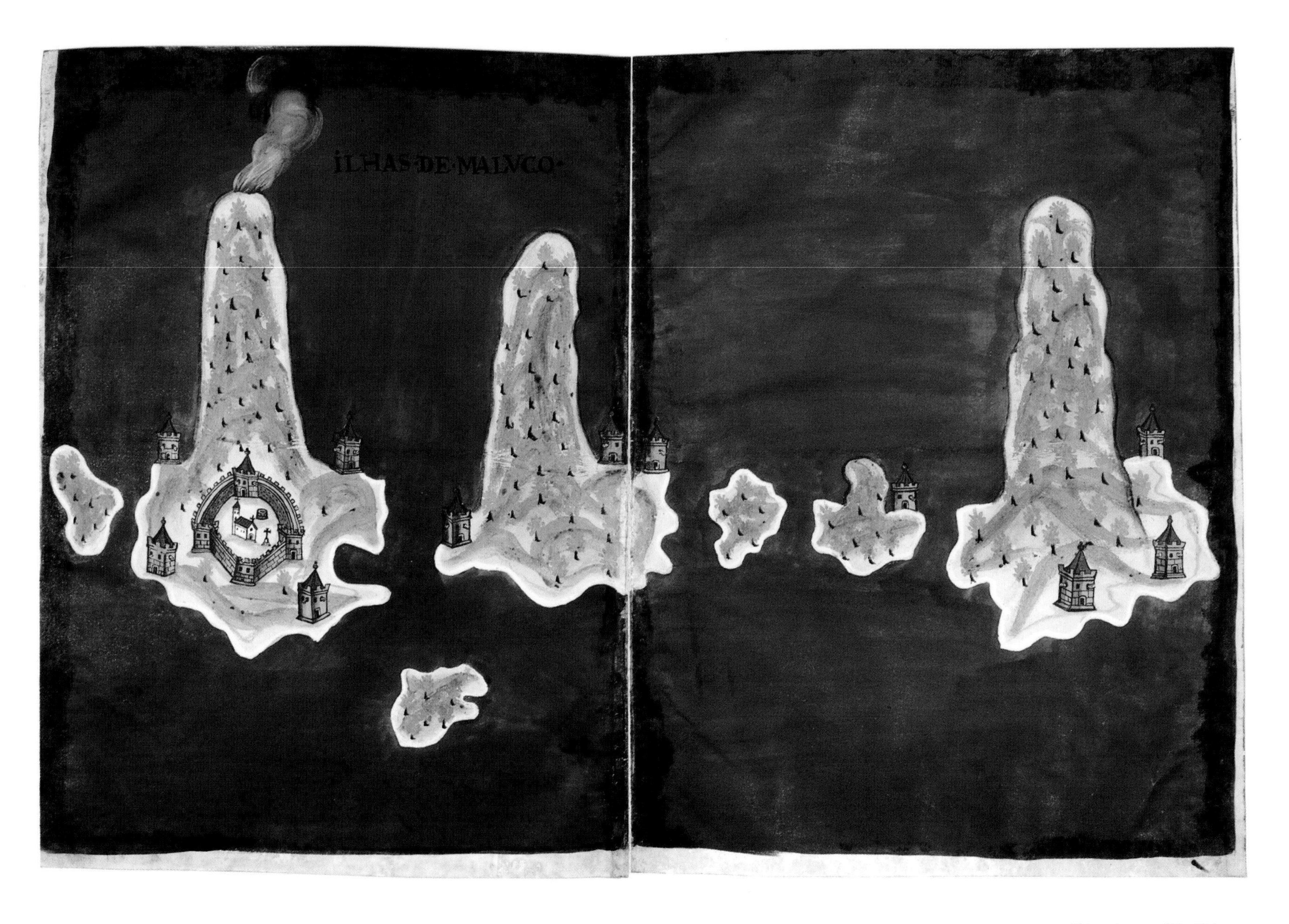

This early map of the Moluccas (probably dating to the early 1500s) illustrates Portuguese strongholds on the precious islands of spice. (Courtesy of the British Museum, London)

THE MOLUCCAS 5

Illness and Inspiration

East of Celébes were the Moluccas—the Spice Islands. Centuries before the European age of exploration, the sultans of Tidore and Ternate were trading spices by sea to Asia. They piled up enormous wealth. In the sixteenth century the first colonizing Westerners in Southeast Asia, the Portuguese, were in the Moluccas, their prime economic drive being to bring spices to Europe. And when in the seventeenth century the Dutch forced the Portuguese out, they monopolized the spice trade.

Wallace sailed from Celebes to the island of Banda, where the Dutch had introduced nutmeg from New Guinea. He saw the old Portuguese fort, and the perfect cone of the island's active volcano—one of many throughout the Moluccas.

From Banda it was a twenty-hour sail to Amboyna, an earthquake island where the Dutch cultivated cloves. The harbor was a revelation. "The clearness of the water afforded me one of the most astonishing and beautiful sights I have ever beheld. The bottom was absolutely hidden by a continuous series of corals, sponges, actiniae, and other marine productions, of magnificent dimensions, varied forms, and brilliant colours. The depth varied from about twenty to fifty feet, and the bottom was very uneven, rocks and chasms, and little hills and valleys, offering a variety of stations for the growth of these animal forests. In and out among them moved numbers of blue and red and yellow fishes, spotted and banded and striped in the most striking manner, while great orange or rosy transparent medusae floated along near the surface. It was a sight to gaze at for hours, and no description can do justice to its surpassing beauty and interest. For once, the reality exceeded the most glowing accounts I had ever read of the wonders of a coral sea. There is perhaps no spot in the world richer in marine productions, corals, shells and fishes, than the harbour of Amboyna."

Fort Belgica (1611), a historical reminder of the many battles fought over the Spice Islands, still silently guards the port of Banda Neira. (Jez O'Hare)

Bottom left: This door graces a colonial building in Banda Neira, the historical center of the spice trade. The earliest records of spice trading date back to A.D. 200, when Chinese of the early Han dynasty took an interest in the Spice Islands. Spanish and Portuguese navigators began arriving in the fifteenth century, and then the Dutch moved in to establish large nutmeg plantations. (Jez O'Hare)

Opposite, clockwise from left: Early engraving depicting the harvest of nutmeg in the famous Spice Islands of the Moluccas (known today as Maluku). The much-prized nutmeg was jealously guarded by local rulers for many centuries. (PhotoBank)

The aromatic nutmeg seed was worth its weight in gold during the heyday of the spice trade in the eighteenth century. (Jez O'Hare)

The scarlet outer covering, or aril, of the nutmeg fruit provides the spice mace. (Jez O'Hare)

In Banda, one of the loveliest and most historically serene of Maluku's islands, nutmeg is still harvested with simple tools. (Jez O'Hare)

The harbor at Ambon (Wallace's Amboyna), the capital city of Maluku (the Moluccas), where Wallace described delicate coral reefs. (PhotoBank)

Right: In the "Realm of the Thousand Islands," as Maluku is known, boat building traditions remain important. (Jez O'Hare)

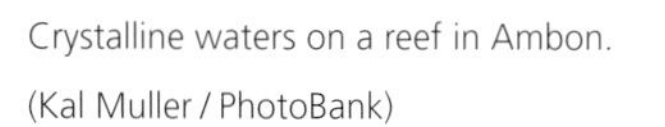

Crystalline waters on a reef in Ambon. (Kal Muller / PhotoBank)

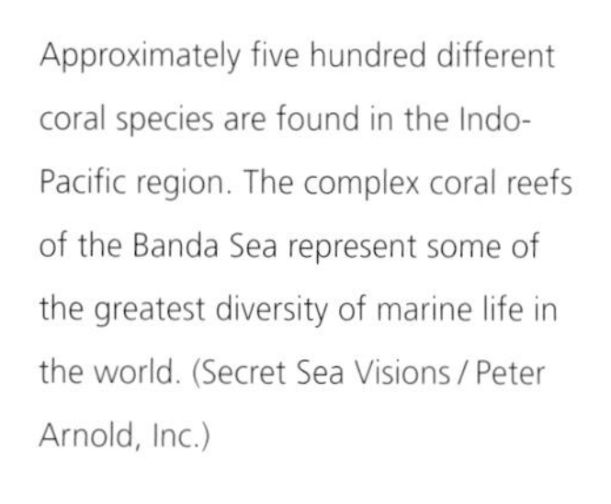

Approximately five hundred different coral species are found in the Indo-Pacific region. The complex coral reefs of the Banda Sea represent some of the greatest diversity of marine life in the world. (Secret Sea Visions / Peter Arnold, Inc.)

By the count of a Dutch ichthyologist, there were almost as many kinds of fish in the waters of this little island as in the waters of all of Europe. Wallace was invited to view a resident's almost perfect collection of fish, preserved in spirits in clear glass jars, along with something like ten thousand shells of a thousand different kinds, arranged in large shallow pith-boxes lined with paper, every specimen fastened with thread.

There were two doctors in the town, both amateur entomologists. One specialized in flies and spiders, with a side interest in butterflies and moths. The other concentrated on beetles. He had lived for years in the islands—Java, Sumatra, Borneo—and in Japan, and everywhere he went he collected. On Amboyna he paid the natives for large and handsome items, and Wallace could not help comparing his own situation with the doctor's: eyes by the hundred, watching over hundreds of square miles, bringing in species that a single collector might not see in ten years.

The peaceful, pristine beaches of many of the tiny islands of Maluku delighted Wallace. (Jez O'Hare)

Opposite: The graceful *Mastigius* jellyfish is a common inhabitant of Indonesia's coastal waters. (Charles Seaborn / Odyssey)

The red lory, one of many exquisite lories found in Maluku. Illegal export of these and other parrots has greatly diminished natural populations. (E. R. Degginger / Animals Animals)

Above right: The brilliant red of the female eclectus parrot contrasts with the chartreuse of her male mate. This species is widely found in Maluku. (Alain Compost)

Wallace went out after birds and collected a crimson lory, a brush-tailed parroquet, a racquet-tailed kingfisher—all Australian types. He ate breadfruit for the first time and was impressed. Plant transfer came to his mind. With speedy steamships and a British invention called the Wardian case, which made it possible to keep plants alive for long stretches at sea, it was "much to be wished that the best varieties of this unequalled vegetable should be introduced into our West India islands, and largely propagated there. As the fruit will keep some time after being gathered, we might then be able to obtain this tropical luxury in Covent Garden Market."

In the Moluccas, Wallace was most at home on the tiny island of Ternate. He had letters of introduction to a Dutchman named Rennesse van Duivenboden, who was "very rich . . . owned half the town, possessed many ships, and above a hundred slaves." Van Duivenboden found Wallace a house, "ruinous" but reparable, and once repaired, quite comfortable, with "ample space and convenience for unpacking, sorting, and arranging my treasures." Wallace used it as a base for three years of traveling in the eastern archipelago. "In this house I spent many happy days. Returning to it after three or four months' absence in some uncivilized region, I enjoyed the unwonted luxuries of milk

A 1724 engraving depicting Ternate, the island from which in 1858 Wallace mailed to Charles Darwin his essay on evolution and natural selection. (Tettoni, Cassio, and Associates PTE Ltd. / PhotoBank)

and fresh bread, and regular supplies of fish and eggs, meat and vegetables, which were often sorely needed to restore my health and energy."

In January 1858 he was away from his home comforts, on Gilolo, a sizeable island a few hours' rowing and sailing from Ternate. Gilolo was perfect Wallace territory. As he wrote to his friend Henry Bates, it was an entomological terra incognita: not a single insect had ever been caught there to be pinned and classified for European science. And Wallace, in the month of his thirty-fifth birthday, was once again in his preferred condition for reflective thinking, as close as he could come to productive solitude, a pondering nineteenth-century Westerner surrounded by wild nature and uncultured man.

Three years earlier, alone in the same way in the wet season in Sarawak, endlessly revolving in his mind what he had read and stored from the works of Alexander von Humboldt, William Swainson, and Charles Lyell, Wallace had conceived the theory that caused Lyell to think harder than he had ever done about the origin of species, and further caused Lyell to communicate this sharpened interest to Charles Darwin. Since then Wallace had continued to ruminate on the fundamental problem still left unresolved: how changes of species were brought about.

On Gilolo, with nothing to do but allow his receptive mind to open itself to ideas, Wallace came down with an attack of malaria. By this stage in his life he was a battlefield veteran of ten years of anopheles mosquito bites, first in the Amazon, then in the islands. Malaria was a recurrent disease. It could come on at any time, hot fits of fever alternating with cold fits of shivering and shaking; and when it struck Wallace he would have to lie down for several hours. With the temperature on Gilolo at 88 degrees Fahrenheit and his body wrapped in blankets, "the problem"—the unsolved question of the origin of species—"again presented itself."

There came to Wallace's malarial mind the grim vision of Thomas Malthus, whose perception of life as an unremitting struggle to survive—populations held in check by disease, famine, and war, the strong surviving, the weak and sick dying—had been a permanent possession of Wallace's thought for close to fifteen years.

Malthus had been talking about human populations. It struck Wallace, lying on Gilolo suffering from his own human sickness and weakness, that the same forces must be even more strongly at work among animals, selecting the strong for survival, the weak for death.

This was a Wallacean flash of light, and in his malarial fever he saw the brilliance of it. He wanted to capture it before it burned out and disappeared like a dream. "In the two hours that elapsed before my ague fit was over I had thought out almost the whole of the theory, and the same evening I sketched the draft of my paper."

Here was Wallace's line of thought: In every animal species, many more individuals were born than would survive. It did not matter how many individuals were born; if the numbers born exceeded the limits set by external constraints, there would be deaths down to those limits. The strongest individuals would be best placed to survive; the weakest would die. And what was true of individuals was also true of species. Those species best equipped to exploit their habitat, find food, and defend themselves against enemies had the best chance of survival. Within species, over time, useful variations would tend to increase; useless or harmful variations would tend to disappear. Superior varieties would ultimately extirpate the original species.

Wallace had formulated the theory of evolution by natural selection.

Tiny atoll islands surrounded by
rich coral reefs abound in Maluku.
(Jez O'Hare)

Over the next two evenings he wrote his paper out in full, for mailing from Ternate, addressed to Charles Darwin.

Wallace's short essay hit Darwin like a thunderbolt. It read like nothing so much as an abstract of what Darwin himself had been working away at for years.

Darwin had hundreds of thousands of words on paper but was still nowhere near completing a publishable book manuscript. Wallace's paper of less than four thousand words made a complete reasoned statement; it could be published as it stood.

If Wallace published first, Darwin would lose the prime value of close to twenty years of his own work. Not money value—there was no money in scientific theory, and in any case Darwin had no money worries. What he would lose was perceived originality, the public establishment of priority in scientific thought by being first in print. For Darwin this was of both personal and professional value, amounting to the principal meaning of his life. As, of course, in its own way it was for Wallace.

Wallace knew essentially nothing about Darwin's thinking on the origin of species, because Darwin had told him nothing. He and Darwin had been writing back and forth, on and off, about various things; but concerning this all-important subject Darwin was not forthcoming.

He was not singling Wallace out for exclusion. For Darwin, the origin of species was a desperately difficult, even dangerous question. He talked and corresponded about it only within a very small circle, and in confidence—lying low.

Within this small circle, Charles Lyell had been urging Darwin to put something into print: "Out with the theory & let it take date—& be cited—& understood." Another friend, the botanist Joseph Hooker, son of the director of the Royal Botanical Garden at Kew, was giving Darwin contrary counsel, advising him to hold off until he could fire his heavy artillery.

Darwin, for his part, was not at all enthusiastic about publishing anything ahead of his big book, however long the book might still take him. "I do not suppose I shall publish under a couple of years," he had told Wallace, in a letter that said nothing substantive about what might be in the book.

Wallace had a book of his own in mind and had done some planning work on it. He was also an immediate and unceasing publisher of articles and essays. But as for the new paper, the one written on Gilolo and mailed to Darwin from Ternate, he had said nothing to Darwin about putting it into print, just that he would be pleased if Darwin would show it to Charles Lyell.

This Darwin did, in deep disturbance at seeing the originality of his life's major work "smashed." And his turmoil was made more dreadful by the fact that, at the moment of his worst imaginable professional crisis, his youngest child, a son less than two years old who had been born retarded, was critically ill with scarlet fever, close to dying.

What to do? Lyell and Hooker came up with a strategy by which the revolutionary ideas of Darwin and Wallace, in all their disconcerting similarity, would be presented jointly at a meeting of the Linnean Society.

This was done very quickly. By calendar and clock, elapsed time from Darwin to Lyell and Hooker and from them to the Linnean Society was much shorter than from Wallace to Darwin. And it was arranged without consulting Wallace.

A meeting of the society was scheduled for the evening of July 1, 1858. The day before, June 30, Lyell and Hooker were able to have the item added to the agenda. In their covering letter they described Darwin and Wallace as "indefatigable naturalists," going on to say: "These gentlemen having, independently and unknown to each other, conceived the same very ingenious theory to account for the appearance and perpetuation of varieties and of specific forms on our planet, may both fairly claim the merit of being original thinkers in this important line of inquiry; but neither of them having published his views, though Mr. Darwin has for many years past been urged by us to do so, and both authors having unreservedly placed their papers in our hands, we think it would best promote the interests of science that a selection from them should be laid before the Linnean Society."

Lyell and Hooker organized the presentation so that the first words to be read were parts of an unpublished essay that Darwin had sketched in 1839 and copied out in 1844, followed by the substance of a letter that Darwin had written to an American botanist in 1857. Wallace's paper came last.

Darwin was not present at the meeting. His little son had just died; he was at home, desolate. Wallace, away on the other side of the world, was not even aware that the meeting had taken place.

Wallace's essay was less than four thousand words. Darwin's two submissions together were shorter than that. But the number of words written by others about the incident runs into the hundreds of thousands. In fact, it has become one of the most closely studied episodes in the history of Western science—and not only because of the paramount importance of the scientific subject matter. The human discontinuities and discordances are as intricate as the scientific continuities and concordances.

What were Lyell and Hooker really up to? Were they genuinely and disinterestedly forwarding the cause of science? Or were they doing Darwin the greatest favor of his professional life, and on the other side of the ledger doing Wallace a huge unmerited disfavor in *his* professional life? And in all this maneuvering, where did Darwin himself stand?

If the paper trail is followed every step of the way, and if the documentary record is read forensically—as evidence, so to speak—a case can be constructed to show what looks like a professional conspiracy by Lyell and Hooker, with Darwin's complicity, to head Wallace off, even to steal his thoughts.

In modern times, a century and more after the fact, this conspiracy case has been made and refuted, resubmitted and rebutted. It remains open, not definitively settled on either side of the balance.

Wallace, for his part, did not think badly of anyone involved. He was habitually generous about other people, and by temperament he was not one to thrust himself forward at anybody else's expense. When, months after the Linnean Society meeting, a letter arrived from halfway around the world about the joint presentation, his response was that he was pleased to have pushed Darwin. This was in fact what happened. Darwin set aside his enormous book, unfinished for so many years, and rapidly wrote a shorter (though still long) one, which was published late in 1859 under the title *On the Origin of Species by Means of Natural Selection, or the Preservation of Favoured Races in the Struggle for Life.* Darwin mentioned Wallace in the introduction, but only briefly. The theory of evolution by natural selection was out in the wide world, and it was publicly, permanently, irreversibly identified with the name of Charles Darwin, not Alfred Russel Wallace.

Marine Biodiversity

For once, the reality exceeded the most glowing account I had ever read of the wonders of a coral sea. There is perhaps no spot in the world richer in marine productions, corals, shells and fishes, than the harbour of Amboyna.

WALLACE, *The Malay Archipelago*

Indonesia's constellation of 17,500 islands is embedded in the shallow waters of the Sunda and Sahel Shelves and scattered throughout the deep Timor, Banda, and Flores Seas. It forms a permeable barrier between the vast Pacific and Indian Oceans and the great landmasses of Asia and Australia. The region is a cauldron of oceanic activity—currents, upwellings, seasonal monsoons—as waters from the Pacific and Indian Oceans attempt to mix and the movements of large equatorial air masses are interrupted by the islands. This array of physical factors, combined with the archipelago's singular geologic history, gave rise to coastal and marine habitats that are among the most productive and species-rich on earth.

At least forty-eight different types of marine ecosystems have been classified, including mudflats and mangroves, seagrass meadows, rare landlocked marine lakes, and coral reefs. Although each type of ecosystem hosts its own bevy of creatures, none compares to coral reefs for sheer numbers of species: 25 percent of all marine species are found in coral reefs, which cover less than 1 percent of the ocean

Top: Extensive mangrove estuaries in the Aru Islands of Maluku. (Jez O'Hare)

Bottom: Stilt roots of *Rhizophora* mangrove provide important refuges for fish fry and shrimp larvae. (Alain Compost)

Fringing reef surrounding Komodo Island. (Jez O'Hare)

Opposite: Coral reefs are among the most productive and species-rich ecosystems found in the archipelago. (Secret Sea Visions / Peter Arnold, Inc.)

floor. Indonesia's reefs, representing one-eighth of the world total, are inhabited by a staggering 2,500 species of dazzlingly colored fish and more than 400 species of stony corals, whose tiny individual polyps form the limestone skeletons that amass over centuries to build the reef's foundations. This impressive scorecard of coral reef species makes the region the world's epitome of marine biodiversity.

Although Wallace was the first Westerner to discover some of the extensive reef areas in eastern Indonesia and to celebrate their beauty in written word, it was Charles Darwin who, in 1842, conducted the first comprehensive study of coral reefs of the world, based largely on his survey of captains' ship logs. Darwin developed a distributional map of reefs and proposed a theory of coral reef evolution and classification.

In Darwin's scenario, a young volcano rising from the ocean depths serves as a substrate to attract free-floating coral larvae. Coral colonies grow to encircle the interface of the volcano and the sea, eventually forming a fringing reef. Through time (measured in millennia), the volcano gradually subsides back into the sea; meanwhile, the reef continues to build upward at roughly the same rate as its land base sinks, forming a barrier reef and enclosing a lagoon that encircles the remaining land. When the entire landmass has completely submerged and the lagoon completely fills the barrier reef, an atoll results. Darwin's simple but elegant theory was finally proven one hundred years after he first proposed it, although his classification of reef types has long been a part of the lexicon of marine biologists.

Fringing reefs, barrier reefs, and atolls are found throughout Indonesia. Each reef habitat is a world that resonates with life, changing according to the influence of currents, tides, sunlight, and moonlight. Fields of staghorn corals with delicately branched spikes, massive boulder corals, and table corals that fan out in rounds as delicate as doilies

seem at first to be subtly hued but lifeless rocks. Closer inspection reveals that the entire surface of the coral is a matrix of hundreds of thousands of tiny, uniformly spaced cups, or calices, each housing its own living animal polyp. At night, the polyps of most coral species expand from their calices to capture food in the water column, exposing brilliantly colored tentacles that transform the stony surface into a lavish profusion of blossoms. Coral polyps augment their diet with the help of symbiotic single-celled plants called zooxanthallae, which are embedded in polyp's tissues and which use light to photosynthesize. In exchange for a protected habitat within the coral calyx, the zooxanthallae produce organic carbon compounds, which enable corals to grow by promoting the calcification process.

Corals form the structure of the reef ecosystem, providing an endless variety of crevices, branches, holes, ledges, and niches for invertebrate and vertebrate dwellers. A soft-bodied nudibranch with vibrant racing stripes of fuchsia, yellow, and purple glides over a coral branch; a commensal goby shrimp arduously removes pebbles from its burrow entrance; clusters of magenta damselfish swirl above the precipice of the reef wall; an enormous grouper rests on a sandy bottom; a shark silently cruises in smooth arcs past the reef wall. Biologists have only just begun to understand the complexity of the coral reef ecosystem, and to identify the thousands of species that inhabit it, even though human societies have reaped its bounties for centuries.

Over one hundred million people now live in the coastal zone of Indonesia. Most of them depend on resources from healthy coral reefs and other marine habitats. With the advent of Indonesia's economic boom, these areas are being swallowed up by conversion and dredging or ravaged by pollution and increased sedimentation from deforested areas. "Over 40 percent of the coral reefs in Indonesia are now in poor condition," laments Soekarno, a senior

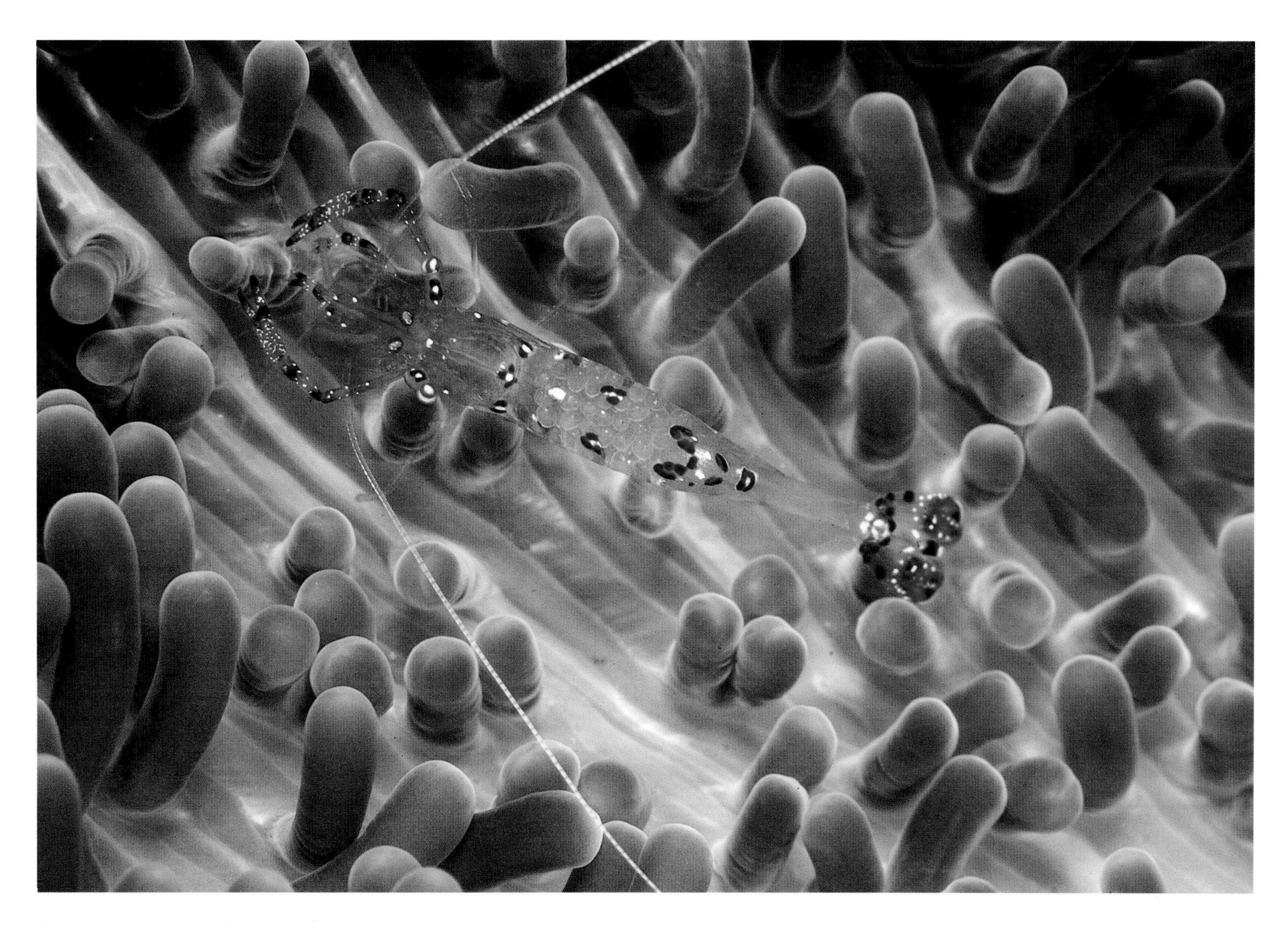

The anemone shrimp, a tiny, delicately colored commensal reef species. (W. Gregory Brown / Animals Animals)

Opposite: Pink anemone fish find protection within their anemone home by using the host's stinging tentacles to fend off intruders. (Secret Sea Visions / Peter Arnold, Inc.)

Showy soft corals and coral grouper. Groupers are popular targets of the live reef fish trade of Southeast Asia that now threatens many reefs in Indonesia. (Kal Muller / PhotoBank)

Above: A pair of lionfish hunting among feather star crinoids that cling to a large barrel sponge. (Fred Bavendam / Peter Arnold, Inc.)

A school of hammerhead sharks, a rare sight of the deeper reaches of remote reefs. (Kal Muller / PhotoBank)

researcher at Indonesia's National Sciences Institute who has studied coral taxonomy for over thirty years. "Unless we improve our management and enforcement capabilities for these precious areas, the majority of our reefs will be lost in thirty to fifty years."

Rili Djohani, a Nature Conservancy marine biologist, has been working in eastern Indonesia to curb another growing threat: destructive fishing practices. "Hundreds of tons of cyanide are squirted and dumped on Indonesia's reefs every year in order to stun large reef fish like groupers and Napoleon wrasse, which are then exported to Hong Kong and China for the lucrative live-fish restaurant business. Smaller fish and the corals themselves are not stunned but killed, leaving behind a dead patch of reef." Djohani says that fishermen who formerly used lines, traps, and nets are now fishing with dynamite—a practice that collects all fish in the area with one blow. "It is a horrible thing to witness, especially when you realize that below the water an entire swath of beautiful reef is being reduced to rubble."

The government of Indonesia, recognizing the economic and biological importance of its marine resources, has enacted fishing restrictions and has established thirty-five marine reserves that currently total 4.6 million hectares. Another 10 million hectares is targeted for marine conservation by the year 2000. The shortage of equipment, facilities, and

Top: Dobo in the Aru Islands, Maluku. About half the population of Indonesia lives in coastal areas and depends on marine resources. (Jez O'Hare)

Bottom: The regal Napoleon wrasse can reach a length of 2 meters. Cyanide fishing for this and other species has led to significant decreases in local populations. (Didier Noiret / PhotoBank)

Marine reserves can afford protection for many species that are currently threatened. *Above:* The massive leatherback turtle must return from the sea to land to lay its eggs, which makes it vulnerable to predation by humans; *left:* shark populations are decreasing because of overharvesting for shark fin soup, a popular Asian dish. (PhotoBank; Jez O'Hare)

Fishing village on Komodo Island.
(Jez O'Hare)

trained enforcement field staff, along with a dearth of information and a lack of awareness and constituency at all levels of society, is crippling these efforts, however. Djohani and others from local conservation organizations work with the government to improve and implement marine park management plans, and they promote sustainable fishing methods among local communities. But instituting change over such a vast area quickly enough to stem the destructive tide is an enormous task—one that determines the fate of Indonesia's reefs.

A source of hope for the reefs may lie in the resurrection of *sasi,* a system of local marine tenure that has been practiced in many fishing communities in eastern Indonesia for centuries. With this system, members of the village maintain exclusive fishing rights to adjacent reefs, often timing and regulating their takes to prevent overharvest. They enforce the right to permit outsiders into their fishing territory—for a price that depends on resource levels gauged by their own fishing takes—or to banish outsiders if conditions are not suitable. "Without such exclusionary rights," notes Bob Johannes, a fisheries expert who has studied traditional fishing communities throughout the Pacific, "the fish one leaves behind can simply be caught by someone else. The most rational course of action, therefore, is to deplete one's resources as quickly as possible. It makes sense to fish with dynamite, even if it destroys the habitat on which fish depend, unless outsiders can be prevented from using dynamite." Johannes advocates official government recognition of local marine tenure traditions: "Where fishing communities possess secure, exclusive rights to fish in an area, it makes sense for them to harvest in a nondestructive manner, for they are assured of all the future benefits."

The pristine reefs in the harbor at Ambon, which Wallace described in 1869, have now succumbed to the ravages of pollution and sedimentation. The same fate may befall remaining reef areas in Indonesia if the call for conservation goes unheard.

Traditional speargun diver of eastern Indonesia. In the past, many coastal villages maintained exclusive fishing rights, thereby preventing overharvest of nearby reef areas. (Rob O'Hare)

Darwin, Wallace, and Precedence

The case for a conspiracy at Wallace's expense surfaced in the late 1960s and was developed in the 1970s and 1980s. Until then, the accepted version of events as set down in published accounts was that those involved in the Linnean Society presentation of 1858—Lyell, Hooker, and Darwin himself—had nothing but benevolent motives: the joint presentation of ideas at the meeting was proffered as evidence of the mutual nobility of Darwin and Wallace, a monument to the natural generosity of both the great biologists. The cognoscenti maintained that Darwin, an eminent scientist, was too much of a gentleman to have countenanced anything ethically unworthy, either as an accessory, allowing Lyell and Hooker to guarantee priority for Darwin's theory, or, far more seriously, as a principal acting covertly and unscrupulously in his own interest, pushing Lyell and Hooker to support him.

The revisionist version—the one charging conspiracy—reopens these awkward issues. David Quammen, reviewing the situation in his book *The Song of the Dodo: Island Biogeography in an Age of Extinctions,* puts the questions in the plainest of terms: Did Charles Darwin cheat and lie in order to protect his proprietary claim? Did he manipulate Lyell and Hooker into helping him? Did he falsify certain documents and destroy others, so as to keep Wallace from what Darwin considered his own rightful glory? If not for the old anointed version's rosy mythologizing, Quammen says, there would be no need for a revisionist version.

Three books, written from differing perspectives, are useful in understanding the elements of the conspiracy theory: Lewis McKinney's *Wallace and*

Natural Selection, published in 1972; Arnold Brackman's *A Delicate Arrangement,* published in 1980; and John Langdon Brooks's *Just before the Origin,* published in 1984. Brooks's examination of the surviving documentary record is the most detailed. He goes into dates on letters; postal delivery times between Ternate and England, and then to Darwin's home at Downe in Kent; obvious gaps in known sequences of correspondence; types of paper and ink; penciled marginal notes; and line-by-line, word-by-word comparisons of text as they appear in Wallace's 1858 paper, Darwin's earlier unpublished manuscripts, and Darwin's later published work.

Brooks decides that Darwin is guilty not only of falsifying documents and putting Lyell and Hooker up to what they did, but of a far greater professional misdeed, the ultimate in scholarly criminality: stealing Wallace's ideas and publishing them as his own. Given the importance of those ideas to Western civilization, this would make Darwin an immoralist on a world scale. Brooks bases his case on chronology. According to what Darwin put on paper at the time, Wallace's communication from Ternate arrived at Darwin's home on June 18, 1858; Darwin sent it on to Charles Lyell (as Wallace in his covering letter had asked him to) the same day. Brooks says that this is not what happened. Given the knowable date of Wallace's mailing leaving Ternate, the known dates of shipping schedules between Ternate and England, and the known British postal schedules between London and Darwin's home, Brooks contends that Darwin would certainly have had Wallace's paper in his hands for as much as two weeks before June 18, probably for as long as a month.

Ch. Darwin

Opposite: Alfred Russel Wallace. (Courtesy of the Natural History Museum, London)

Charles Darwin. (Mary Evans Picture Library)

Brooks's revisionist chronology raises interesting questions. If Darwin, opening his mail on June 18, indeed found Wallace's thunderbolt paper, which Darwin saw as smashing twenty years of his own work, it must have been only a few hours before he sealed it again and sent it to Lyell, as Wallace had requested, in the outward mail of the same date. Long enough, certainly, to read it and be stunned, but hardly long enough to assimilate all the implications, professional and personal. Perhaps long enough, though, to copy out its fewer than four thousand words before sending the original to Lyell? On the other hand, if Darwin was not telling Lyell the truth about when he received Wallace's communication—if in fact Darwin had had it for anything up to a month before June 18—why did he lie? And why did he keep it for so long before sending it on?

In essence, so the conspiracy theory goes, he sat on it to give himself time to figure out a self-serving strategy. In this reading, what Darwin decided was, first, to allow Lyell and Hooker to do some apparently high-minded but actually less than honorable work for him at the Linnean Society; and second, to take Wallace's ideas, especially on the topic of divergence—a crucial aspect of the theory of evolution by natural selection—and work them into his own manuscript version of the theory, without attributing them to Wallace but rather claiming them as his own. In short, stealing them.

Brooks's case, for all its forensic detail, remains circumstantial. This is inevitably so because vital pieces of paper, known to have existed, are missing: first in importance, Wallace's covering letter and envelope; next, six other letters that Wallace is known to have written to Darwin in the same period; and letters of Lyell and Hooker from the time when the Linnean Society presentation was arranged.

Brooks and the other revisionists assign great significance to these gaps in the documentary record, most especially because, on the evidence of all the rest of Darwin's extensive lifelong correspondence files, he was a systematic keeper, not a discarder. If there is no "smoking manuscript," there is nevertheless what smells like a lingering whiff of smoke rising from burned paper.

Barbara Beddall, another close student of the subject, lists, in three scrupulously considered scholarly articles (1968, 1972, and 1988a), the documentary record available for comprehending the situation: some fifty items. Her listing makes the gaps glaringly obvious. Her quietly understated inference is that someone sanitized the files. Who was the sanitizer? Darwin's son Francis, who after his father's death published an edition of his letters? Or Darwin himself? No one knows. Short of the rediscovery of the missing documents, no one will ever know the true nature of the crime, or even whether any crime was committed.

Among specialists, James Moore, co-author of a biography of Darwin and a student of Wallace's life, says: "In the spring of 1858 Wallace knew nothing that Darwin didn't know already, so there was nothing for Darwin to 'steal' from him, except possibly glory. There was undoubtedly a 'delicate arrangement' in getting natural selection into print with Darwin's name first and without Wallace's permission. But there was no rip-off." And others argue that although Wallace laid out the theory in his four-thousand-word paper, it was Darwin who, drawing on years of research, presented the compelling evidence.

Jonathan Hodge, who wrote a doctoral dissertation on Darwin and Wallace, maintains that no "serious rip-off thesis can be confirmed by the evidence. It remains a remarkable case of convergence on two theories which are very nearly the same." Hodge contends, however, that Darwin, again by virtue of the weight of his years of research, is intellectually entitled to be considered the senior partner.

This is the generally accepted verdict, even though the circumstances of 1858 are murky. But who, truly, was first? Where does precedence for the theory of evolution by natural selection lie, within the strictures of scientific publication? And what if Wallace, instead of sending his 1858 paper to Darwin, had sent it directly to a journal in London to be published? That was his customary practice with his many papers from the islands. It was what he had done with his 1855 paper from Sarawak, the one that came so close to cracking the theoretical nut of evolution, the one that piqued the interest of Charles Lyell, the one that Lyell drew to Darwin's attention. If Wallace had followed the same direct-to-publication route with his 1858 paper—the decisive one, the revolutionary one—he would have been in print ahead of Darwin. Unambiguously and irreversibly, he would have had priority. He would have been first. No matter what Darwin published, and no matter when, Darwin would have been second.

Ernst Mayr, a distinguished evolutionary biologist and historian of science who rejects the conspiracy theory, says that "in the competition between these two, Wallace behaved quite admirably. He knew that Darwin had worked on this problem for more than twenty years and did not claim priority even though he was the first to have completed a publishable manuscript. Presumably, whenever one speaks of natural selection, one ought to mention both Darwin and Wallace."

Yet that is not how history has treated Wallace. History has Darwin first and preeminent, whereas Wallace is perceived (if he is seen at all) as secondary and subsidiary, a far lesser light— "Darwin's Moon," as he is dubbed in the title of a short biography. Consider the several high-powered biographies of Darwin. There is nothing comparable for Wallace, a lack that at least partly reflects the perception of who came first. Without question, on his own merits Wallace is worth a major biography, not just as a scientist but as a figure of his times. As Jonathan Hodge notes, "We want him text and context, as a Victorian whose life has lines leading outward from himself in so many fascinating directions."

Jane Camerini, an insightful historian of science with a grounding in zoology, geography, and cartography, argues that Wallace deserves to be seen not just as a disembodied thinker, a brain on paper, but as a man worth knowing for his Victorian goodness: a combination of modesty, amiability, and intelligence; an inherent optimist, not a brooder, not one to dwell on failure; a man capable of good relations, someone whose being and presence enhanced life for others. "He was a moral man," she says, "true to his own thinking and wide-ranging beliefs, from organic evolution to spiritualism and socialism."

James Moore, however, sees warts as well: "Wallace was awkward, angular, self-opinionated; a perpetual outsider who did not suffer fools gladly, who called a spade a spade, regardless of his social class. He lacked a self-critical spirit, and I know of no lesson he ever, frankly, unlearned." And while he energetically opened new scientific avenues, he appeared just as energetically to pursue any number of intellectual dead-ends.

But Mayr, whose own pioneering ornithological expeditions in New Guinea beginning in the 1920s were very much in the mold of Wallace, whose life in biological fieldwork and research has been as long as Wallace's, contends that we would not be doing justice to Wallace to treat him only as a scientist: "He was exceptionally sensitive to the welfare of other human beings, and throughout his life he was aware of the beauty and, I am tempted to say, the sacredness of living nature."

Darwin seated opposite Charles Lyell and Joseph Hooker (standing). (The Royal College of Surgeons, London / The Bridgeman Art Library).

6

The isolated Kei Islands, where Wallace first encountered the flora, fauna, and people of the easternmost reaches of the archipelago. (Jez O'Hare)

Sand spit in the Kei Islands, where some of the longest and most pristine beaches in the world are found. Wallace spent a few days in these islands and marveled at the "dazzling whiteness" of the sands and the water "transparent as crystal." (Jez O'Hare)

THE ISLANDS OF KÉ, ARU, AND NEW GUINEA 6

Land of the Bird of Paradise

When, unknown to Wallace, his words were being read in London in second place after Darwin's, Wallace himself was where he had always wanted above all to be: in bird of paradise country, away at the eastern end of the Malay Archipelago.

He had taken his first voyage in that direction the year before, in 1857, sailing to the Aru Islands on the kind of trading vessel that had been used for centuries to make the journey, a Bugis prau, with a Javanese-Dutch captain and a crew mostly Macassarese, the rest miscellaneous. From Celebes to Aru was a thousand miles, the prau's top speed was five knots, and Wallace said he never enjoyed twenty days at sea more in his life. He had his own little cabin, six and a half feet long (just five inches more than he measured lying down), five and a half feet wide (less than his arm span), and four feet high (two feet one inch less than he measured standing up). He laid his mattress on the floor of split bamboo covered with fine Macassarese cane mats, and hung his long-barreled guns, revolver, and hunting knife from the thatch roof. He had his insect boxes, lamp, and books, his canteen, his little store of luxuries, rice and curry for dinner, and his own people to wait on him—his permanent companion, Ali, and two others he had hired at Macassar, named Baderoon and Baso. He liked the smell of the prau as much as he hated the smell of steamships—all that modern paint, tar, grease, oil, and varnish. For him, breathing the aroma of vegetable fibers recalled quiet scenes in the green and shady forest. "I was," he wrote, "much delighted with the trip, and was inclined to rate the luxuries of the semi-barbarous prau as surpassing those of the most magnificent screw-steamer, that highest product of our civilization."

The farther east Wallace sailed, the farther he was voyaging into the unfamiliar. His last landfall before Aru was at the Ké Islands, to him an entirely new world. "But few

Above: The faces of Aru Islanders reflect the mixing of western and eastern Indonesian ethnic stocks. (Jez O'Hare)

Glassy surface of the sea in a protected inlet of the Aru Islands. Wallace sailed among these small islands in local fishing vessels. (Jez O'Hare)

These elegant praus are still hand-built from wood using a traditional technique in which the hull is often honed from a single tree. Wallace used small vessels like these to travel short distances in the Aru Islands. (Jez O'Hare)

The coconut crab, the world's largest terrestrial invertebrate, can attain weights of more than 3 kilograms on the isolated islands of Aru. This relative of the hermit crab is a nocturnal omnivore that forages opportunistically on animal and vegetable matter, contrary to traditional stories that describe it as a coconut predator. (Jean-Paul Ferrero / Auscape)

Right: Because they are prized for their sweet, delicately smoke-flavored flesh, coconut crabs, once numerous on the islands of the eastern archipelago, are becoming rarer and are now legally protected. (Jean-Paul Ferrero / Auscape)

European feet had ever trodden the shores I gazed upon; its plants, and animals, and men were alike almost unknown." Sea cliffs, weather-worn limestone pinnacles, forested uplands, little bays and inlets, beaches of dazzling whiteness, the sea as calm as a lake, and the glorious sun of the tropics throwing a flood of golden light over all—to Wallace "the scene was . . . inexpressibly delightful"; he could "dream of the wonderful productions hid in those rocky forests, and in those azure abysses." Ashore for the better part of a week, he sighted all kinds of butterflies and other insects, "among the most precious and rare in the cabinets of Europe."

Finally, early in January, the prau made Aru, well before the opening of the trading season. The little coastal settlement of Dobbo was lifeless. The Aru people told Wallace he was ahead of time as well for the bird of paradise, at least the species he knew as *Paradisaea apoda*. From what he could gather, it did not come into full plumage until September or October. The villagers brought him skins, but they were so badly preserved and dirty that they looked like last year's unsaleables; Wallace would not buy them. The second species he had heard about as coming from Aru, the king bird of paradise, he did not sight at all, not even a skin.

There was no shortage of other curiosities of natural life, however: corals and sponges and sea slugs thrown up on the beach after windy nights; hermit crabs in their commandeered seashells on the floor of the forest, even up in the trees; and yellow-spotted spiders with two-inch bodies, spinning their webs for Wallace to run into while he was chasing butterflies. On his first day of collecting he came in with some thirty species.

A few days later he caught "one of the most magnificent insects the world contains," a great birdwing butterfly, *Ornithoptera poseidon*. The experience sent him into a collector's ecstasy. "I trembled with excitement as I saw it coming majestically towards me, and could hardly believe I had really succeeded in my stroke till I had taken it out of the net and was gazing, lost in admiration, at the velvet black and brilliant green of its wings, seven inches across, its golden body, and crimson breast." He had seen birdwings in cabinets in England, but it was "quite another thing to capture such one's self—to feel it struggling between one's fingers, and to gaze upon its fresh and living beauty, a bright gem shining out amid the silent gloom of a dark and tangled forest. The village of Dobbo held that evening at least one contented man."

Still, he was more than discontented not to be collecting birds of paradise. Dobbo was offshore of the principal Aru island of Wokan. Wallace wanted to go there and hunt in the interior. He was organizing the use of a boat and putting together a crew when a scare shut everything down. Pirates were about, plundering and killing at sea. It was no time to be on the water. The scare faded, but the weather, which had been bad, continued to get worse. In the end it was two months before Wallace could sail for Wokan.

Wallace marveled at the radiant splendor of birdwing butterflies, which can have wingspans of over 25 centimeters. This species, *Ornithoptera goliath,* is found in Irian Jaya, the center of evolution for the birdwings. (Alain Compost)

The first few days there were wet. Collecting was poor; Wallace was sighting next to nothing. Then Baderoon came back from shooting with a specimen "which repaid me for months of delay and expectation."

It was the first bird of paradise Wallace had seen with his own eyes in its native habitat—a complete bird, not just a skin. It was not alive, but at least it was only recently dead. Wallace gave it his fully focused field naturalist's attention. "It was a small bird, a little less than a thrush. The greater part of its plumage was of an intense cinnabar red, with a gloss as of spun glass. On the head the feathers became short and velvety, and shaded into rich orange. Beneath, from the breast downwards, was pure white, with the softness and gloss of silk, and across the breast a band of deep metallic green separated this colour from the red of the throat. Above each eye was a round spot of the same metallic green; the bill was yellow, and the feet and legs were of a fine cobalt blue, strikingly contrasting with all the other parts of the body. Merely in arrangement of colours and texture of plumage this little bird was a gem of the first water; yet these comprised only half its strange beauty. Springing from each side of the breast, and ordinarily lying concealed under the wings, were little tufts of greyish feathers about two inches long, and each terminated by a broad band of intense emerald green. These plumes can be raised at the will of the bird, and spread out into a pair of elegant fans when the wings are elevated. But this is not the only ornament. The two middle feathers of the tail are in the form of slender wires about five inches long, and which diverge in a beautiful double curve. About half an inch of the end of this wire is webbed on the outer side only, and coloured of a fine metallic green, and being curled spirally inwards form a pair of elegant glittering buttons, hanging five inches below the body, and the same distance apart. These two ornaments, the breast fans and the spiral tipped tail wires, are altogether unique, not occurring on any other species of the eight thousand different birds that are known to exist upon the earth; and, combined with the most exquisite beauty of plumage, render this one of the most perfectly lovely of the many lovely productions of nature."

Wallace had collected the king bird of paradise, *Cicinnurus regius*. That was one. *Paradisaea apoda* would make two. At the end of a week, Ali came in "triumphant" with a fine male specimen. "The ornamental plumes had not yet attained their full growth, but the richness of their glossy orange colouring, and the exquisite delicacy of the loosely waving feathers, were unsurpassable."

Wallace found a tree with a number of birds high up in the thickest of the foliage. They were flying and jumping about, so he could not see them well, but eventually he got in a successful shot. His first kill was a young bird, and as things turned out he shot nearly a dozen young ones before he so much as sighted an adult.

The male king bird of paradise displays for a potential mate with acrobatic feats. This was one of the few species encountered by Wallace. (Alain Compost)

Superb marksmen, hunters in the eastern islands of the archipelago still use bow and arrow to hunt birds, as did the hunters employed by Wallace to collect birds of paradise. (Alain Compost)

The Aru hunters did their shooting of birds of paradise with a bow, using arrows with a conical wooden cap that killed by force of impact, leaving no wound, shedding no blood. The place for hunting was a tree where the male birds assembled in season, their plumage fully developed, to do their mating display before the females. "They open their wings," Wallace wrote, "stretch out their necks, shake their bodies, and keep the long golden plumes opened and vibrating—constantly changing their positions, flying across and across each other from branch to branch, and appearing proud of their activity and beauty." The long downy side feathers "are erected vertically over the back from under and behind the wing, and there opened and spread out in a fan-like mass, completely overshadowing the whole bird. The effect of this is inexpressibly beautiful. The large, ungainly legs are no longer a deformity—as the bird crouches upon them, the dark brown body and wings form but a central support to the splendour above, from which more brilliant colours would distract our attention,—while the pale yellow head, swelling throat of rich metallic green, and bright golden eye, give vivacity and life to the whole figure. Above rise the intensely-shining, orange-coloured plumes, richly marked with a stripe of deep red, and opening out with the most perfect regularity into broad, waving feathers of airy down,—every filament which terminates them distinct, yet waving and curving and closing upon each other with the vibratory motion the bird gives them; while the two immensely long filaments of the tail hang in graceful curves below."

With two species of the bird of paradise in hand, it was Wallace's season to do some

The Raggiana bird of paradise, a close relative of the species encountered by Wallace in the Aru Islands. Here, a male vigorously displays to a female. During the elaborate mating ritual, the male tries to entice a potential mate with fluttering wings held high and dancelike movements of the head. (Bruce Beehler)

displaying of his own. "Rejoice with me," he wrote to Samuel Stevens, "for I have found what I sought; one grand hope in my visit to Aru is realized: I have got the birds of Paradise (that announcement deserves a line of itself); one is the common species of commerce, the Paradisea apoda; all the native specimens I have seen are miserable, and cannot possibly be mounted; mine are magnificent"—the best ever. And he knew he was the first Westerner in history to describe the display of the bird of paradise for other Westerners. "I have discovered their true attitude when displaying their plumes, which I believe is quite new information; they are then so beautiful and grand that, when mounted to represent it, they will make glorious specimens for show-cases, and I am sure will be in demand by stuffers. . . . The other species is the king bird (Paradisea regia, Linn.), the smallest of the paradisians, but a perfect gem for beauty; of this I doubt that any really fine specimens are known, for I think Lesson only got them from the natives; I have a few specimens absolutely perfect." He gave in to some preening self-congratulation: "I believe I am the only Englishman who has ever shot and skinned (and ate) birds of Paradise, and the first European who has done so alive, and at his own risk and expense; and I deserve to reap the reward, if any reward is to be reaped by the exploring collector."

His shooters were coming in with a bird of paradise a day, and he was offering the Aru hunters incentives for more. He gave them a first flask of arrack free; the second they had to pay for, and for each one after that they had to bring in a bird of paradise. Other kinds of birds as well were everywhere for the collecting. "At early morn, before the sun has risen, we hear a loud cry of 'Wawk—wawk—wawk, wok—wok—wok,' which resounds through the forest, changing its direction continually. This is the Great Bird of Paradise going to seek his breakfast. Others soon follow his example; lories and parroquets cry shrilly, cockatoos scream; king-hunters croak and bark; and the various smaller birds chirp and whistle their morning song. As I lie listening to these interesting sounds, I realize my position as the first European who has ever lived for months together in the Aru Islands. . . . I think how many besides myself have longed to reach these almost fairy realms, and to see with their own eyes the many wonderful and beautiful things which I am daily encountering. But now Ali and Baderoon are up and getting ready their guns and ammunition, and little Baso has his fire lighted and is boiling my coffee, and I remember that I had a black cockatoo brought in last night, which I must skin immediately, and so I jump up and begin my day's work very happily."

Life was good. At night, though, the dogs of Aru invaded Wallace's open hut. Once he left two *Paradisaea apoda* skins on his table, wrapped in paper; in the morning they were gone, nothing left but a scattering of feathers. Another time Ali had just finished skinning a king bird; he dropped the skin, a dog seized it, and before Ali could get it back it was torn to tatters. Those Aru dogs—they would drink the lamp oil and eat the wick.

All kinds of ants attacked Wallace's specimens; arsenic meant nothing to them. There were centipedes and millipedes everywhere, little scorpions, mosquitoes, and sand flies worse than mosquitoes. Wallace was bitten nonstop for a month, until his feet broke out in inflamed ulcers. He could not walk, which meant he could not collect. He would crawl to the river to bathe, see a wonderful butterfly, a blue-winged *Papilio ulysses*, and not be able to chase it down—for him a collector's punishment worse than ulcers.

Over a stretch of six weeks, Wallace was laid up more than half the time, and he could not be sure how soon he would have the use of his legs again. His bird and insect boxes were full, his stores nearly exhausted. He called it a day, distributed what salt and tobacco he had, gave the owner of his hut a flask of arrack, and headed back to Dobbo.

At the height of the trading season, a hundred small boats were in the harbor and fifteen praus were hauled up on the beach. Heaps of dried trepang were being bagged; bundles of mother-of-pearl shells, tortoise shells, and edible birds' nests were being exchanged for English calico and American unbleached cotton, crockery, tobacco, guns, and arrack. There was heavy drinking, much singing in the night, and betting on cockfights. There were deaths from fever. Baderoon quit because Wallace scolded him for laziness. Ali, who had been at Wanumbai on the main island for a fortnight, came back with sixteen more bird of paradise skins, glorious specimens, paid for in silver dollars.

It was time to leave, ahead of the monsoon coming out of the east. The fifteen praus sailed away in convoy to the west.

Another place for birds of paradise—*the* place, so Wallace was ready to believe as an article of faith—was the north coast of New Guinea, centering on the little trading village of Dorey and a scattering of offshore islands.

For a white man this was wild territory, even further off the map of European experience than Aru. Very few Westerners had ever been there. But Dorey was where the French naturalist René Lesson became the first white man in history to be able to put on paper the ecstatic experience of seeing a living bird of paradise in flight. Plumes trailing, a brilliant meteor—Lesson had stood so transfixed by the vision that he forgot to shoot. And this vision was vouchsafed to him on a walk of no more than a hundred yards into the forest. In just a few weeks, Lesson shot twenty birds (noting in passing that in fact they did have feet); and the natives brought him more skins. Lesson left with more than forty, so many that when he got back to Paris he could afford to give them away casually.

The turkey-sized crown pigeons of Irian Jaya are the largest pigeons in the world. (Alain Compost)

For Wallace, Dorey shone like the promised land. While he was at Dobbo, traders and rajahs told him he could travel safely there and beyond, and they gave him the names of places where he would find what he was looking for. "Their accounts excited me so much that I could think of nothing else."

Wallace sailed to Dorey from Ternate on a trading schooner, with Ali and three other helpers, "in full confidence of success," and from the very first ran into nothing but disappointment and distress. He did not get in a single shot at a single bird of paradise. His hunters brought in only one, of a common species, and the Dorey people did not encourage him to believe that he would be able to get more. They turned out not to be hunters themselves; they did not even preserve skins.

Going after insects, Wallace spiked his ankle and developed an ulcer and an inflammation that had to be leeched and lanced and poulticed for weeks, leaving him housebound on crutches, in naturalist's purgatory again. His hunters got sick too.

To make things worse, a Dutch war steamer came to Dorey on a surveying expedition, and the villagers took everything tradeable to the ship, including food. Wallace was left short, down to making two meals out of a single small parroquet. One of the survey's draftsmen was an amateur naturalist. He showed Wallace a pair of skins of a very rare crown pigeon that he had bought on an island not far away. This was Wallace's first and only sight of the species, and he could not get his hands on the specimens—more frustration.

He kept hearing from the villagers that the real place for birds of paradise was a hundred miles away, somewhere called Amberbaki. He sent Ali there. Ali came back saying that he saw nothing but the same kinds as at Dorey, only scarcer; and that, in any case, bird of paradise skins came not from Amberbaki itself but from the interior, beyond several mountain ridges—farther and farther out of reach.

When Wallace was fit enough again he went out after insects and was able to collect dozens of species a day, on his best day ninety-five distinct kinds, in five hours of searching among dead leaves, beating foliage, and hunting under rotten bark, followed by six hours pinning and setting out specimens and separating the species. Altogether, in three months of working over a patch not much more than a mile square, he collected more than a thousand distinct kinds of beetles. It was his best beetling ever.

That was a consolation prize. The overwhelming brand left on his memory by Dorey was of ants swarming his worktable, carrying off his insects, even tearing them loose from his gummed cards, or getting in his hair and crawling over his face, his hands, his whole body, in his bed at night. He was never free of them. When he set out bird skins to dry, blowflies swarmed on them and laid masses of eggs that turned into maggots. When he was not sick himself, Ali and his other men were; one died of dysentery.

Wallace was used up. He wanted to be away from Dorey as much as he ever wanted to go there. When the schooner came back, he was happy to put the place behind him.

The island of Batchian in the Moluccas was better in every way. Wallace sailed there from Ternate and stayed more than five months. The very first day that Ali went out to shoot, he came back with trophies hanging from his belt. "Look here, sir," he said, "what a curious bird." What he held out was a puzzle to Wallace. "I saw a bird with a mass of splendid green feathers on its breast, elongated into two glittering tufts; but what I could not understand was a pair of long white feathers, which stuck straight out from each shoulder. Ali assured me that the bird stuck them out this way itself, when fluttering its wings, and that they had remained so without his touching them. I now saw that I had got a great prize, no less than a completely new form of the Bird of Paradise."

Straight away Wallace wrote to Stevens: "Here I have been as yet only five days . . . I believe I have already the finest and most wonderful bird in the island. I had a good mind to keep it secret, but I cannot resist telling you. I have a new Bird of Paradise! of a new genus!! quite unlike anything yet known, very curious and very handsome!!! . . . I consider it the greatest discovery I have yet made."

One, two, three exclamation points—this was as excited as Wallace ever got in his life over a bird. He sent his shooters out for more, prepared the skins, shipped them back to England, and had the honor of having the British Museum formally name the new species after him: *Semioptera wallacii,* Wallace's standard-wing.

His months on Batchian brought him to another high point of collecting experience. During his very first walk into the forest he saw an immense butterfly, dark in color, with white and yellow spots. It was a female of a new species of birdwing, "the pride of the Eastern tropics." He could not catch it; it was on a leaf out of his reach. For two months he kept his eyes peeled but only made two more sightings, a female and a male. He was beginning to despair. Then, after he saw one hovering over a flowering shrub and missed it because it was too quick for him, he went back the next day and caught a female. A male would be the big prize. The next day he caught one: "a perfectly new and most magnificent species, and one of the most gorgeously coloured butterflies in the world. Fine specimens of the male are more than seven inches across the wings, which are velvety black and fiery orange. . . . The beauty and brilliancy of this insect are indescribable, and none but a naturalist can understand the intense excitement I experienced when I at length captured it. On taking it out of my net and opening the glorious wings, my heart began to beat violently, the blood rushed to my head, and I felt much more like fainting than I have done when in the apprehension of immediate death.

One of the magnificent birdwing butterflies of eastern Indonesia. This species, *Ornithoptera croesus,* was first discovered by Wallace on the Maluku island of Batchian. (Alain Compost)

I had a headache the rest of the day, so great was the excitement." He named the species *Ornithoptera croesus.*

He had been planning to leave Batchian, but the lure of more great birdwings changed his mind. One of his men cleared bushes and creepers away from the flowering shrub for him, and he ate lunch there regularly, catching a birdwing a day. Birdwings had also been seen along a stream. Wallace tried collecting there, but the wet stones were too slippery for him. He put a man on extra wages to do the catching and got one a day, on good days two or three. He accumulated more than a hundred.

Ornithoptera croesus and *Semioptera wallacii* were great prizes, and he was collecting new birds too. A fine racquet-tailed kingfisher. A splendid deep blue roller. A lovely golden-capped sunbird. And a real rarity, the Nicobar pigeon, a ground-feeding bird, therefore neither in theory nor in real-world daily life a high flyer, that nonetheless was found on widely separated islands all along the archipelago, which meant that it had somehow adapted to flying long distances over water.

A bonus for Wallace on Batchian was good health. For two months he was free of fever, for the first time in two years. The worst things that happened to him were minor: an earthquake shock that rattled his little house for five minutes and shook the

The Nicobar pigeon is found throughout the archipelago and is easily recognized by its luxuriantly long neck feathers. (PhotoBank)

trees like a gust of wind, and a robbery in which he had his collection boxes disturbed and his cash box and the keys to it stolen.

The thief was a convict serving his time at hard labor. Wallace was told by the commandant of the fort that if he caught the man in the act he could shoot him. As things turned out, Wallace did sight the robber again, on the government boat that took him, his helpers, and all his valuable specimens away from Batchian. It was a four-tonner with outriggers, a mat sail, and twenty tomtom-beating, betel nut–chewing, cigarette-smoking rowers, courtesy of the sultan of Ternate. For seven days Wallace shared a little thatched cabin with a Chinese trader, a schoolmaster's wife, a servant, some Javanese soldiers, a stowaway poisonous snake, and two time-expired convicts, one of them the robber. The thief did not have Wallace's cash box, and Wallace did not shoot him. In his time on Batchian, Wallace shot only birds, some opossums, and a civet cat; and on the boat he killed only the snake.

Wallace's last voyage in search of the bird of paradise was to the island of Waigiou, just west of New Guinea. It took him forty days to get there from Ternate. Along the way, on Goram, he had to buy a boat, a little prau, repair it, and put a crew together. He paid them in advance, and they repaid him by deserting. He recruited new men, but then ran into contrary winds and currents, making headway more by rowing than sailing. Off a coral islet, the prau was carried away by a strong current, leaving two men stranded ashore, marooned for weeks, subsisting on roots, leaves, and shellfish. Nearing Waigiou at last, the prau was eight days among reefs, and on the reef five times in twenty-four hours.

While Wallace was battling along offshore, he could hear birds of paradise screaming inland. René Lesson had been on Waigiou, and according to what he wrote he had bought several species there. After all Wallace's time and trouble, this was the sort of return he was looking to match, even surpass.

But the Waigiou people told him there was only one kind on the island. He saw some and identified the species as the red bird of paradise—rare, in fact found only on Waigiou, but not new: it was known to Lesson. Wallace shot at males one after another but brought none down.

There was no house for him to use. He and his men built a long shed, with the mat sails of the boat for walls. The site he picked was by a tall fig tree. Birds of paradise were fruit eaters, and when the figs ripened they came. One morning while Wallace was drinking his coffee, a male settled on the treetop. Wallace picked up his gun, but he could not get a clear shot. It took him several days of watching and one or two misses before he managed to kill a male red bird in the most magnificent plumage: the head, back, and shoulders a rich yellow, the throat a deep metallic green, with little erectile crests on the forehead, side plumes rich red with delicate white points, and for middle tail feathers "two long rigid glossy ribands, which are black, thin, and semi-cylindrical, and droop gracefully in a spiral curve."

Wallace shot another male out of the tree. After that the birds stopped coming. A month of miserable weather went by, and he shifted operations to Bessir village, where he had been told there were men who were skilled at catching birds of paradise.

The house he moved into with the agreement of the local chief was eight feet square, with only five feet of headroom at the ridgepole of the upper space, where he slept, and less underneath, where he worked. "Here I lived pretty comfortably for six weeks, taking all my meals and doing all my work at my little table, to and from which I had to creep in a semi-horizontal position a dozen times a day; and, after a few severe knocks on the head by suddenly rising from my chair, learnt to accommodate myself to circumstances."

The Bessir bird of paradise hunters were specialists, the only ones on the island. They did not use a bow and arrow like the Aru people. Their way was to make a noose on the end of a long cord, bait it with fruit, set it up in a tree where the birds congregated, and wait underneath, dawn to dusk, for as long as it took, which might be days.

They had not dealt with a white man before; Wallace was the first ever in their village. He negotiated in sign language, showed them hatchets, knives, beads, and handkerchiefs, set a price, paid them in advance, and waited.

The hunters brought in birds as promised. Wallace was struck by their honesty. But they did not have his collector's requirements clear in their minds. The first bird was alive, but tied up in a small bag, its wing and tail feathers crushed. Others were snared one by one a long way away, and the hunters, rather than walk miles to Wallace with a single bird and then out again, stockpiled them for a week or ten days, tied up and hanging by the leg. Of course the birds struggled and damaged themselves, or were left hanging so long that the tied leg swelled and started to putrefy, or they died hanging.

When birds were brought in alive and well, seven or eight of them, Wallace tried to keep them that way, in a bamboo cage. He fed them fruit, boiled rice, grasshoppers, and other insects. The first day they ate greedily and were all activity. On the second day they ate but were less energetic. On the third day he almost always found them dead at the bottom of the cage, some of them falling off the perch in convulsions.

As for Wallace, he was suffering from what he called brow-ague, an intense pain in his right temple, as bad as the worst toothache, every day, right after breakfast, for a couple of hours. And then fever. He could not eat properly, what little there was to eat. All his white man's food was gone, and Waigiou was not a natural larder. He and his men were having to get by on meager rations, rice and sago, tough cockatoo and pigeon, to the cusp of starving—ferns and the tops of pumpkin plants, wild tomatoes only the size of gooseberries, and a few green pawpaws. In the hardest of hard times the Bessir people ate a fleshy seaweed, boiled. Wallace could not stomach it; it was too salty and bitter for him. He had two tins of soup set aside for dire emergencies, and that was his lifesaver.

He had managed to do some useful collecting: seventy-three species of birds, including many very rare and twelve entirely new; beautiful butterflies, including a superb green birdwing; and a whole series of red birds of paradise, twenty-four fine specimens. Now the monsoon was coming on. If he was going to get back to Ternate, he needed to sail ahead of it.

The spanker and jib of his little prau had been stored in the roof of the shed-house he had built, and rats had nested in them and gnawed them through in twenty places. At sea, the prau ran into contrary winds and currents and squalls, and could make next to no headway. In the salt air and the glaring sun, Wallace's lips burned terribly—painful, bleeding at a touch and not healing. A wind sprang up, rising toward hurricane force, the worst the old Bugis steersman had ever been through; it had the old man calling upon Allah. Allah was kind, the wind died away, and the prau made port safely at Ternate.

"Thus," wrote Wallace, "ended my search after these beautiful birds." His words echo with wistful unfulfillment. The great collecting moments that should have been his with birds of paradise had never materialized. He never experienced an ecstatic moment of encounter and triumph of the kind that he had with birdwing butterflies. At Aru, his first-ever bird of paradise in the hand was not from his own gun; Baderoon was the one who shot it and brought it to him. And his first bird of paradise of an entirely new species, the one that was formally named for him, was not his own sighting and shooting either, it was Ali's. In identifying this new species on Batchian, Wallace had extended the known range of birds of paradise to the Moluccas. That was a first, but no more than a modestly useful addition to scientific information. Then too, as birds of paradise went, Wallace's standard-wing was not spectacularly beautiful or even unusually big. Stuffed and mounted, perched in a cabinet beside a *Paradisaea apoda,* it would look diminished, diminutive, even dull.

Wallace's sharpest pang was over the great unexplored territory of New Guinea: "The vast extent of country east of long. 136d is quite unknown; but there can be little doubt that it contains other and perhaps yet more wonderful forms of this beautiful group of birds."

Wallace had observed in the Amazon how location determined differentiation of species. For birds of paradise, New Guinea turned out to be a demonstration case. A cordillera a thousand miles long divided the island into biological regions on either side, one tropical, the other subtropical; and between the cordillera and the coastal ranges to the north was a low-lying area. On the south side of the Baiyer-Wahgi Pass, the flank plumes of birds of paradise were red; on the other side they were orange-yellow. On mountains on either side of a valley only twelve miles wide were species of the six-wired bird of paradise as different from each other as if they had evolved separately on oceanic islands.

Wallace never saw any of this. He could eventually list in his mind eighteen known species of the bird of paradise, of which he himself had managed to sight only a handful. There turned out to be forty-two species, almost all of them in New Guinea. The last was not positively identified until 1938, more than a century and a half after Linnaeus formally named the first species for science, and more than three-quarters of a century after Wallace collected his single new species. Not until the final quarter of the

A

B

Because of its great altitudinal gradient, from sea level to almost 5,000 meters, the large island of New Guinea harbors a spectacular variety of ecosystems. As in Wallace's time, much of the island remains unexplored. (a) The vast mangroves of Bintuni Bay in Irian Jaya support a wide diversity of life, including fish, shrimp, birds, and reptiles. (b) A small village nestles in a high valley amid the upland rain forests of Irian Jaya. (c) Jaya Peak in Irian Jaya, at 4,884 meters the highest elevation between the Himalayas and the Andes, has rare equatorial glacial meadows. (Jez O'Hare; Bryan and Cherry Alexander; Kal Muller / PhotoBank)

A

B

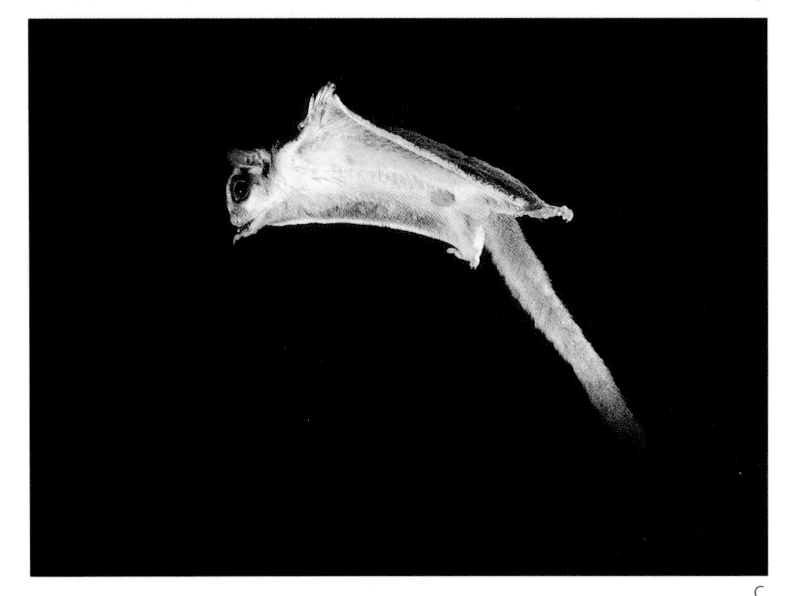

C

Some of the curious denizens of Irian Jaya and Papua New Guinea that Wallace might have seen had he been able to venture into the island's interior: (a) Bennett's cassowary, one of three cassowary species found on the island. This enormous, flightless bird—one of the largest land animals on New Guinea—feeds primarily on fallen fruit. (b) The spotted cuscus, prized by Irianese hunters for its meat and thick, woolly fur. (c) The honey glider, a marsupial that, like the flying squirrels of western Indonesia, has adapted to life in the forest canopy. The patagium, a flap of skin that connects its fore and hind limbs, allows it to glide gracefully from tree to tree in search of flower nectar. (PhotoBank; Alain Compost; Alain Compost)

(d) Goodfellow's tree kangaroo, an endemic species of the montane forest of the central cordillera. In the absence of monkeys, tree kangaroos have taken over the niches of arboreal fruit and leaf eaters on this island. (e) The young of the green tree python of Irian Jaya are a vivid yellow. (f) Brightly colored caterpillars feast on ginger blossoms in the montane forest. (g) The exotically fashioned blossoming stem of the Irian Jaya orchid. (Jean-Paul Ferrero / Auscape; Alain Compost; Alain Compost; Alain Compost)

D

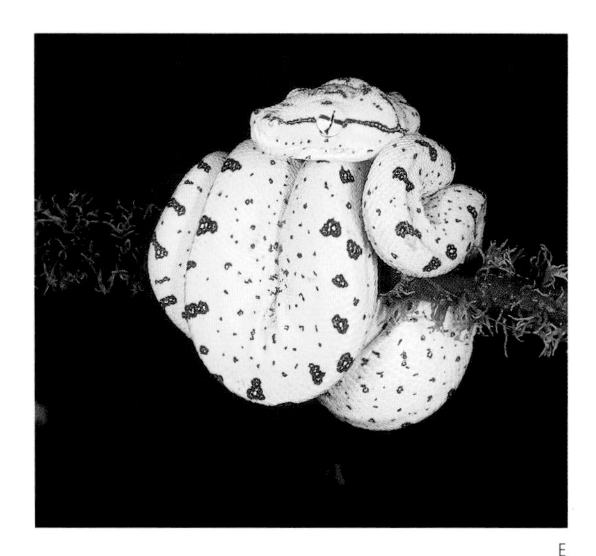

E

F

G

twentieth century would a naturalist have information complete enough to map the range of each species and be able to write from first-hand experience, as Thomas Gilliard did, about the astonishing variety and extravagance of the panoply of plumage of the whole bird of paradise family: skirts, whips, lacelike feathers, capes, twisted enamel-like wires, erectible expandable fans, saber-shaped tails, patched or mirrorlike iridescent plumage, jade and opal mouths, nutlike and leaflike wattles, naked garishly painted areas of skin. And of the strange and beautiful movements of the males in their displays: some dancing on the ground, freezing, then spinning with their circular feather skirt extended ballerina-like; others dancing in columns of skylight that they let in through the roof of the forest by laboriously stripping away leaves; others hanging from the limbs of their chosen display trees in shimmering pendulous masses.

Wallace in his own exploring and collecting was a hundred years too early for this overflowing ecstatic naturalist's experience. But the century of bird of paradise studies that came after him unfolded as a classic display of his fundamental and formative theories of biogeography.

Ecology and Behavior of Birds of Paradise

A

Fascinated by the improbable fans, wires, shields, elongated streamers, and extravagantly colored plumage of the birds of paradise, several nineteenth-century artists attempted to record nature's eccentricities. (a) The superb bird of paradise by John Gould (1804–1881) and William M. Hart (1830–1908); (b) the Marquis de Raggio bird of paradise by Gould; (c) the King of Saxony bird of paradise by Hart. (The Natural History Museum, London / The Bridgeman Art Library)

In such a country, and among such a people, are found these wonderful productions of Nature, the Birds of Paradise, whose exquisite beauty of form and colour, and strange developments of plumage are calculated to excite the wonder and admiration of the most civilized and the most intellectual of mankind, and to furnish inexhaustible materials for study to the naturalist, and for speculation to the philosopher.

WALLACE, *The Malay Archipelago*

With their flamboyant array of plumes, wattles, breastplates, and strange feather bauble accoutrements, birds of paradise of the island of New Guinea (today politically divided into the Indonesian province of Irian Jaya and the independent country of Papua New Guinea) have long captivated the imagination of Westerners and local people alike. Early Malay traders called them *manuk dewata,* "God's birds." The first accounts of birds of paradise in Europe were from the 1520s, when skins were brought back from the region by Portuguese traders who called them *pássaros do sol,* or "birds of the sun." Because the bird corpses were often sold with their feet amputated to accentuate only the splendid feathers, early European explorers thought they had no feet, and described them as heavenly birds who could never touch the ground. The myth was perpetuated by Linnaeus, who in 1758 named the greater bird of paradise *Paradisaea apoda,* or the "footless bird of paradise." When Wallace began his quest for birds of paradise in 1854, only fourteen species had been described. After many frustrating attempts, he discovered only one new species (Wallace's standard-wing, *Semioptera wallacii*); however, he was the first Westerner to see and describe the bizarre behavior of the greater bird of paradise in the wild.

B

C

Precipitous mountains serve as effective barriers, isolating highland valley species throughout the island of New Guinea. (Alain Compost / Peter Arnold, Inc.)

Opposite: A resplendent headdress of bird of paradise feathers, worn by a New Guinea highlander warrior. (Bruce Beehler)

The forty-two species of bird of paradise now known to science are classified in the family Paradisaeidae and dwell primarily in the highland area of New Guinea (the six exceptions are four Australian species and two single-species genera in Maluku). They are generally heavy-bodied birds and, though strong fliers, are reluctant to fly long distances or cross large expanses of water. The many deep valleys and steep ridges of New Guinea serve as effective barriers, allowing evolution of these species to occur in relative isolation from each other. This accounts for the diversity of species in such a circumscribed area.

Birds of paradise diverged from their closest relatives, the crows, around 30 million years ago. But what evolutionary forces led to the bewildering diversity of forms and behaviors found in this group of birds? Why, in some species, are there drab-colored females and outrageously plumed males, so unlike each other as to be mistaken for different species, while the males and females of other species are nearly identical in their cryptic coloration? Why do some species bear strong, short, thick bills and others long, graceful, recurved ones? And what would account for the spectrum of mating systems, from stable pair bonds to utter promiscuity, found in this group of species? It was not until 1978, when a young ornithologist named Bruce Beehler ventured to New Guinea to conduct his research for a degree from Princeton University, that the evolutionary complexities began to be revealed. "My doctoral study focused on four species that shared a tract of montane forest. Each exhibited a distinct mating system, and I attempted to correlate these behaviors to foraging ecology."

Birds of paradise are generally omnivorous, feeding on fruits and insects, but the proportion of each type of food varies with the species and also the life stage of the individual. Beehler found that the variation in diet affects the foraging range of an individual, which in turn affects the social organization, sexual behavior, and morphology of a particular species. Bill shape provided the major clues to preferred food types. For example, in the monogamous species he studied (in which a male and female pair-bond and remain together for an entire breeding season or for life), males and females resemble each other and have drab coloration that blends into their forest surroundings. They feed on small fruit, readily available but unpredictable throughout the year and providing water and carbohydrates but little protein. Bill shape in these species tends to be strong and stout, much like that of a crow, facilitating manipulation of fruits. In species feeding on these nutritionally poor food sources, it takes two parents to find enough food to feed nestlings, and selection has favored a monogamous mating system in which both male and female pair-bond and cooperate to raise a brood.

Beehler also discovered that species that feed on protein-rich fruits (such as nutmegs) and insects, harder to find but nutritional bonanzas, were generally polygynous—one male had many female mates. Bills in these species come in various forms, from sickle-shaped to short and broad, reflecting the wider range of food types the birds utilize. Because of the protein-rich food sources, females can raise nestlings without the help of the male, emancipating the male from parental duties and allowing him to be as promiscuous as possible. In these species, selection has favored male characteristics and behaviors that increase the attractiveness of the male to the female. The result is extreme sexual dimorphism—marked differences in the physical characteristics of male and female of the same species.

The most profound example of this is the Raggiana bird of paradise. About the size of a small crow, the male is adorned with a pale gold crown, emerald chin and throat, a lush brown breast cushion, and a radiant cascading spray of orange pectoral feathers almost twice as long as its body, contrasting sharply with the dull brown and olive-colored female. Males congregate in traditional display arenas called leks, which they visit year after year, arriving daily during the breeding season to begin calling early in the morning in hopes of attracting females to the site.

Beehler describes the astonishing display: "They advertise their presence to females within earshot by calling loudly: 'wau, wau, wau, *Wau, Wau, Wau,* WAU, WAU, WAU, WAU,' with increasing volume and speed. If a female responds by joining them in the display tree, the males will initiate a courtship dance: they raise their orange cape plumes, shake their wings, and hop from side to side, while continuing to call. After a brief bout of noisy display behavior, the males become silent and lean upside down from their perches, with their wings thrown forward; their erect orange plumes form a resplendent fountain of color. The group holds this pose until the female, who moves silently among them, selects a mate and crouches beside him. The other males watch passively as the chosen male performs a precopulatory dance and then mounts and mates with the female. The female separates from her mate soon thereafter and flies off to her nest, where in a day or two she will lay an egg."

The European plume trade, which late in Wallace's lifetime threatened the most spectacularly feathered species, has all but died out since the late nineteenth century. In Indonesia and Papua New Guinea, all birds of paradise are now officially protected by laws, which, however, are difficult to enforce. Hunting of birds of paradise for traditional costumes and ceremonies is still common practice, but perhaps the biggest threat to their survival is loss of their forest habitat. Beehler calls birds of paradise "flagships for biodiversity. They attract attention and are very important forest seed dispersers. The future of New Guinea's forests depends on them, and by creating reserves for birds of paradise, we may be able to save many other species as well."

7

Komodo Island, east of Java.
(Jez O'Hare)

Perfectly shaped volcanic cones rise through the clouds over eastern Java. Ancient volcanic soils on Java and Bali permit intensive agriculture, unlike the nutrient-poor soils of most of the non-volcanic islands of Indonesia. (Lindsay Hebberd / Woodfin Camp & Associates)

7 JAVA, SUMATRA, AND HOME

Realizations

In the wake of his exhausting and disappointing expedition to Waigiou, Wallace began to speak in his letters about being weary. He was in his late thirties, not as resilient physically as he used to be, especially after so much sickness and near-starvation. "I cannot," he wrote, "now put up so well with fatigue and privation." He had been in the islands twice as long as he had been in the Amazon, and he was ready to think of going home.

In his final months in the archipelago he was back again in the west, this time on the big islands of Java and Sumatra.

Java had more active volcanoes than any place of its size in the world; spectacular ruins of ancient cities; terraced agriculture; a big botanical garden; orchids in profusion in the mountains; peacocks, which Wallace collected; tigers, which he did not; and rhinoceroses, which he never saw.

The Sumatran tiger, the largest predator in Indonesia, is now threatened with extinction because of habitat loss and overhunting. (PhotoBank)

Sumatra had orangutans, which he never saw either: their range was limited, and he was never in their part of the island. Sumatra also had an extraordinary butterfly, *Kallima paralekta*, which he had difficulty seeing even when it was right before his eyes. In flight it was richly colored, very conspicuous. At rest, perched on a twig with wings folded, it looked exactly like a dead leaf—mimicry, a perfect protective adaptation, which Wallace took great pains to observe and describe.

He saw flying lemurs and leaping langurs in the trees. He came upon elephant tusks on the forest floor, but no elephants in the flesh; on Sumatra they were in retreat, deprived of habitat by advancing cultivation. As for rhinoceroses, he did not contemplate shooting a specimen and boiling it down; he acquired a cranium and some teeth from natives.

His hunters brought in a male hornbill, a sizeable bird with a big curved beak. They told him they had shot it while it was feeding its mate, which was shut up in a hole in

Few people have glimpsed the shy Sumatran rhinoceros, now very rare throughout its forest habitat. (Original drawing by W. Jardine, courtesy of The Natural History Museum, London)

a tree. Wallace had read about this extraordinary avian behavior. The female and her egg were inside the hole, which was plastered up with mud, and the male brought food all through incubation and the birth and fledging of the young. Wallace being Wallace, of course he had to go and observe and collect. "After crossing a stream and a bog, we found a large tree leaning over some water, and on its lower side, at a height of about twenty feet, appeared a small hole, and what looked like a quantity of mud, which I was assured had been used in stopping up the large hole. After a while we heard the harsh cry of a bird inside, and could see the white extremity of its beak put out. I offered a rupee to any one who would go up and get out the bird, with the egg or young one; but they all declared it was too difficult, and they were afraid to try. I therefore very reluctantly came away. In about an hour afterwards, much to my surprise, a tremendous loud hoarse screaming was heard and the bird was brought me, together with a young one which had been found in the hole. This was a most curious object, as large as a pigeon, but without a particle of plumage on any part of it. It was exceedingly plump and soft, and with a semi-transparent skin, so that it looked more like a bag of jelly, with head and feet stuck on, than like a real bird."

A

Birds of the Javan rain forest: (a) the rare endemic Javan hawk eagle, (b) blue-eared kingfisher, and (c) crimson sunbird. (Alain Compost)

B

C

A

B

C

Mimicry, whereby species cryptically blend with their surroundings to evade predators or entice prey, or take on the coloration and behaviors of less vulnerable model species, attests to the forces of natural selection. The English naturalist Henry W. Bates, Wallace's friend and traveling companion in the Amazon, was the first to clearly define the concept of mimicry. (a) The leaf insect resembles nearby leaves with its details of veins and lichen spots; (b) a walking stick resembles rotting leaves on the forest floor; (c) a horned frog sits among leaf litter. (PhotoBank; Tim Laman; Frans Lanting / Minden Pictures)

D

E

F

Katydids (d) on moss-covered bark, (e) on a banana leaf, and (f) in a head-standing threat position mimicking a predatory wasp. (Frans Lanting / Minden Pictures; Frans Lanting / Minden Pictures; Kjell Sandved / Photo Researchers, Inc.)

The banded leaf monkey, or langur, feeds on the fruits, flowers, and leaves of the forest canopy in Sumatra. (PhotoBank)

Herds of wild Asian elephants still roam the forests of Sumatra, but they are increasingly threatened by habitat loss and conflicts with man. (PhotoBank)

When some natives caught a siamang, a kind of gibbon, a small tailless ape, Wallace bought it and put it on a leash long enough for it to hang from some poles he set up under his verandah. The siamang would let his Malay helpers play with it, but it did not like Wallace. He tried to win the creature over by feeding it himself, but one day it bit him. "I lost patience and gave it rather a severe beating, which I regretted afterwards, as from that time it disliked me more than ever." He took it with him to Singapore. It was the first siamang ever seen there. He wanted to bring it back to England, but it died.

The siamang is the largest of the gibbons and is found only in the lowland forest of Sumatra. Using a pouch under the chin as a resonator, adult pairs sing loud and complicated duets to establish and maintain their territories. (Tom McHugh / Photo Researchers, Inc.)

At Singapore all sorts of captive living things were for sale, from crickets to songbirds to pythons to tigers to women. Wallace, waiting for his ship to sail in February 1862, was not actively in the market. But then he saw a trader with two lesser birds of paradise. Live birds of paradise were not a common public sight in Singapore; Chinese merchants in the town or rich Indians from Calcutta were always ready to buy them for their private menageries. These two birds were in a cage five or six feet square. They seemed to be in good health. Wallace priced them; the man was asking a hundred pounds. That was extortionate, but Wallace could not resist. For him it was a chance at a historic first—to be, of all men, the one to bring birds of paradise to Europe alive.

On Batchian, Wallace had had birds of paradise in a cage, with thoughts of taking them home. They died within three days. These two at Singapore must have survived much longer in captivity, having been weeks at sea en route from the eastern archipelago. That had to be a good sign. Even so, from Singapore to London, a voyage halfway around the world, was going to take as long as six weeks, out of the tropics into a cold climate in winter. What were the chances that Wallace could get the birds to London alive?

At least he did not think food was a serious problem. Birds of paradise were not finicky eaters; they liked all manner of things, seeds and fruits and insects. So Wallace had in mind a shipboard diet of bananas and cockroaches. Bananas were plentiful at Singapore, and in his eight years in the islands no sea transport he had ever taken was short of cockroaches.

Such were the best-laid plans of the world's greatest field naturalist. What he had not taken into account was that he was booked on a steamer of the Peninsular & Orient line, meaning the highest modern standard of respectable British cleanliness. Meaning that cockroaches were scarce—and on the passenger decks, next to nonexistent. Meaning that the greatest scientific insect collector in the world had to repair every evening to the storeroom in the forecastle, set traps, and brush roaches into a biscuit tin, managing only a few dozen a night, hardly enough for a single meal for his two birds.

At Bombay, where Wallace's ship docked for three or four days, the passengers went ashore to a hotel. Wallace brought his birds to the verandah, and they were a great attraction. Between Bombay and Suez they continued to do well. Construction of the Suez Canal had been begun but was not finished. Wallace and his birds would be making their crossing out of the East to Alexandria, the entry point of the Western world, by rail. It was February, cold, almost frosty. Wallace did not want his tropical birds freezing. The railway officials were difficult. He had to insist on the rarity and value of the birds to get them to agree to put the cage in a box-trunk.

Out on the Mediterranean it was colder still; Wallace was a long way from banana territory, and again his ship was clean of cockroaches. He was worried about keeping the birds' strength up. His only recourse was to break his voyage for a fortnight to go insect-hunting one last time, on one last island. For numbers of insect species, Malta was not Dorey, but Wallace found what he needed at a bakery—cockroaches unlimited. He stocked up and set off for Marseilles.

There he had more trouble with railway officials before he got permission for the birds to travel in a guard's van and for himself to go in and feed them.

The crossing of France was a night ride north—further into the cold of the late European winter. The birds did not seem to suffer.

At last, the English Channel, and finally, finally, British soil. Both birds arrived in perfect health. Twenty-four hours after they reached London, they were at the gardens of the Zoological Society in Regent's Park.

They were put on public show in a room in the museum building, in a wire cage twenty feet by eleven. They were given boiled rice, bread, fruit, and vegetables, and there was no shortage of London cockroaches. They liked them. They did not like each other. They were both males, and even though there was no female to compete for, they could not stand the sight of each other's displays and had to be separated, on opposite sides of a screen.

One lived for a year, the other for two. Wallace thought longer life was possible. "I feel sure that if a good-sized conservatory could be devoted to them, or if they could be turned loose in the tropical department of the Crystal Palace or the Great Palm House at Kew, they would live in this country for many years."

First to last, Wallace had shipped back from the islands the amazing total of 125,660 specimens: 310 mammals, 100 reptiles, 8,050 birds, 7,500 shells, 13,100 lepidopterans,

83,200 coleopterans, 13,400 other insects. His two living birds of paradise had gone to Regent's Park. Orangutan skeletons and skins had been sold off to private collectors and museums, his late adopted baby and all. Samuel Stevens had been conscientiously disposing of the rest. Wallace kept for himself nearly three thousand bird skins of a thousand species, at least twenty thousand beetles and butterflies of about seven thousand species, and some quadrupeds and land shells.

He spent months sorting them, and he could have settled for being what he called a "regular species monger." There was fascination in that "humble occupation." But he had "wider and more varied interests," for example geological time, the glacial period, physical geography, zoological geography, anthropology, sociology, and special applications of natural selection. To name but a few.

For someone who had classified himself when young as constitutionally lazy, who said he always had to be pushed to do anything, Wallace's mental activity and productivity in maturity were prodigious. By age twenty-nine he had had two books in print, and that was before he went to the islands, where he wrote more than forty papers. Back in England, over the rest of his life he wrote twenty-two more books. And he kept turning out shorter pieces all the time—"shorter" by his measure meaning anything up to about two hundred pages. In his last fifty-five years he put something into print on average every six weeks. His thoughts appeared in nearly two hundred publications. When a twentieth-century scholar did some bibliographic arithmetic on Wallace, he counted 753 titles, ten thousand printed pages, millions of words.

Wallace's long book about his eight years in the islands was published in March 1869 as *The Malay Archipelago: The Land of the Orang-Utan and the Bird of Paradise—A Narrative of Travel, with Studies of Man and Nature.*

The reviews were excellent. From the beginning it was recognized as a classic. It went on in Wallace's lifetime to be reprinted fifteen times in England and translated all around the world. It was Joseph Conrad's favorite book.

Wallace dedicated it to Charles Darwin, "not only as a token of personal esteem and friendship, but also to express my deep admiration for his genius and his works." Wallace had brought back a gift for Darwin from the islands, a wild honeycomb from Timor. When he went to visit Darwin at his country home in Kent, Darwin set up a white stone in the garden to mark the occasion, as he did for other visiting great minds.

In London, Wallace met Charles Lyell and Thomas Huxley. He was not as shy as he had been. He discussed the nature of life with them. He did not, however, have any talent or tolerance for small talk or for empty conviviality; parties bored him. With Henry

Si Munjon Coal Works nr. Sarawak
May 8th. 1855.

My dear Sir

I should have written before to acknowledge the receipt of your letter containing your report on my Singapore Curculionidae, and thank you for the trouble you took. I assure you it was most interesting & agreeable to me to find that so many of my insects were new, and has given me much encouragement to persevere in my search after the smaller species. You will already probably have heard through Mr Stevens that I have found another good locality here, which I continue to work very hard at. Curculionidae altogether do not bear quite so large a proportion to the other families as at Singapore, neither are the Anthribidae so numerous among them, but there is still a great number of fine things. Of Cerastonema I have two – three species all I think different from the beautiful one which you were so kind as to name after me. Mecocerus Gazella is very abundant, and there

Parkstone, Dorset.
Septr. 23rd. 1901

…r Fowler

I am afraid
… you much information
… you require; but
…at many tropical birds
…s, especially the larger
… Drongos (Edolius &c) &
… hawks & buzzards, in
…s or stomachs I
… found & elytra a
…n. I think among the
… of the American Board
…e you will find
… detailed accounts of
… of the examination of

Old Orchard,
Broadstone,
Wimborne.

…r Waterhouse

Can you give me
…t available estimate of
… number of described, or
…leoptera – or those in
…ume?
… also be much obliged
…ll obtain from Mr. Kirby
…r is now in charge of the
…a) the approximate number
… in that order.
…any estimate has recently
…e – the number of all
…nsects.
…nly want them for the

Bates he went to visit Herbert Spencer, the social theorist who actually coined the phrase "survival of the fittest." Wallace, the perpetual student of territory and behavior, took note of Spencer's method of dealing with the problem of maximizing the life of the mind in a world not filled with working minds. Spencer chose to live in a boarding house, deliberately among commonplace people, so as to avoid the mental excitement of too much interesting conversation. He was an insomniac, and an evening's tedium before bed improved his chances of sleeping, allowing him to conserve his mental energy for his days of thinking.

Wallace's own way of keeping the wearying world at the right distance was to leave London and go and live in the country. At forty-two he married. His wife, Annie Mitten, was the daughter of a botanist, still in her teens. Wallace in the islands, living the life of a single field naturalist, moved scores of times in eight years. At home, in almost five decades as a family man, he changed habitations more often than the standard Englishman of his class and time but orders of magnitude less often than in the islands, on average only once about every five or six years, mostly within a radius of about a hundred miles from London—not altogether out of touch with the world of science in the city but far enough away to be left alone.

He was finished with expeditions to collect specimens. He kept on collecting and arranging facts. And in his county-bound domesticity his mind continued to range, as it always had, around the world and backward and forward in time: "An accurate knowledge of any group of birds or insects, and their geographical distribution, may assist us to map out the islands and continents of a former epoch. . . . By the collection of such minute facts alone can we hope to fill up a great gap in the past history of the earth as revealed by geology, and obtain some indications of the existence of those ancient lands which now lie buried beneath the ocean. . . . It is for such inquiries that the modern naturalist collects his materials; it is for this that he still wants to add to the apparently boundless treasures of our national museums, and will never rest satisfied as long as the native country, the geographic distribution, and the amount of variation of any living thing remains imperfectly known. He looks upon every species of animal and plant now living as the individual letters that go to make up one of the volumes of our earth's history."

The Amazon and the Rio Negro had shown Wallace the importance of the geographical distribution of palm trees and fishes and monkeys, and by extension of all plants and animals. There was no better place on earth than the Malay Archipelago for the full burgeoning of this approach. Archipelago-wide, there was the most striking evidence of the dynamism of the physical earth—volcanoes, earthquakes, signs of changes

A portrait of Alfred Russel Wallace and letters written during his travels. (Courtesy of The Natural History Museum, London)

Gunung Api, or "fire mountain," a relatively young volcano in the Banda Islands of Maluku. (Jez O'Hare)

in sea level, islands that must be the remnants of continents. On a given island, so many varieties and species of a given animal. From one island to the next, such striking differences between biological communities. And so much to think about concerning the distribution of species.

Detailed observation of the physical and natural world had formed the substrate of Wallace's intellectual hunger to stay for years in the islands. For many more years after he came home, he wrote at great length and in great depth about the connections between natural history research in the present and the history of the earth. In 1876 he published a book in two volumes entitled *The Geographic Distribution of Animals;* in 1878 another entitled *Tropical Nature and Other Essays;* and in 1880 another entitled *Island Life.* For sweep of thought, for ability to incorporate masses of detail into worldwide synthesis, no one surpassed Wallace. He was the father of the modern science of biogeography.

Top: The Bornean crested lizard can change its color rapidly to blend in with its surroundings. *Bottom:* Young Malay civet peers out from the safety of a tree hole. (Frans Lanting / Minden Pictures)

Of all Wallace's books, *The Malay Archipelago* was the only one that ever made him any money. Not that he was interested in piling up money for the sake of it. That was not what his life was about. From the islands he had written to his brother-in-law: "Now, though I always liked surveying, I like collecting better, and I could never now give my whole mind to any work apart from the study to which I have devoted my life. . . . The majority of mankind are enthusiasts only in one thing—in money-getting; and these call others enthusiasts as a term of reproach because they think there is something in the world better than money-getting. It strikes me that the power or capability of a man in getting rich is in an *inverse* proportion to his reflective powers and in *direct* proportion to his impudence. It is perhaps good to *be* rich, but not to *get* rich, or to be always trying to get rich, and few men are less fitted to get rich, if they did try, than myself."

For a number of years he lived on the sale of his specimens, and on the income from investments Samuel Stevens made for him. If he had stayed with that portfolio, he and his family would have been all right. Instead he took the advice of friends, who steered him into riskier things, all of which turned out so badly that by his mid-fifties he was sliding into real money trouble.

He had applied for paying jobs—museum director, supervisor of Epping Forest—but was never appointed. Charles Darwin always did much better with money than Wallace: he came into the world with plenty, and turned it into a fortune in the stock market. When Wallace was in his late fifties and worrisomely low in funds, Darwin and others recommended him for a civil list pension from the British government, as official recognition of his scientific achievements. It was enough for modest comfort, for life.

Wallace did not pension his mind. He kept on thinking day and night. (Like Herbert Spencer he was not a good sleeper and often lay awake cogitating.) And he wrote nonstop, about everything under the sun and in the heavens above. He considered and rejected the possibility of life on Mars. He was interested in a plan for a model of the globe, to be housed in a building six hundred feet high with viewing stairs and ladders. He had suggestions for museums with exhibits laid out not by taxonomic classifications but by regions. He had views on public education. He had theories about the racial classifications of humankind and about the origins of language. He thought there were good reasons why women should have a better life, with more personal freedom and independence. He was for vegetarianism in principle (though not in practice), against vivisection, against vaccination. He never gave up on phrenology, and became certain there was deep truth in spiritualism.

Along the way he parted intellectual company with Darwin on the question of whether the laws of evolution by natural selection that applied so universally to plants and animals also applied in all their rigor to human beings. Wallace decided not. The difference lay in the human mind. Animals had to adapt physically or die. Humans, though, could make all sorts of mental adaptations, and not just for physical survival, but to open the possibility of social and cultural change in the direction of the better, including the moral and spiritual better.

Wallace called one of his books *The Wonderful Century*. He thought the nineteenth century was indeed wonderful in many ways, but not always morally. He was something of a social anthropologist, observing and trying to understand tribes of men as well as species of animals. He had had experience of both Civilized Man and Savage, and he did not think either kind of human had a monopoly on goodness or even good sense. He was impressed by the way in which "primitive" peoples could maintain their societies without an oppressive weight of state apparatus, police, prisons, and armies—in short, the reverse of Europe—but he was still ready to believe that the Savage would be benefited by intelligent colonization. At the same time, he had deep misgivings about his own civilization, especially its economic workings. He had radical views on poverty and injustice. He wanted to see universal competition replaced by universal cooperation. He favored the nationalization of land. He declared himself a socialist.

He did not seek conventional honors any more than he coveted money. Honors came to him as and when others decided: the founders' medal of the Royal Geographical Society, the gold medal of the Société de Géographie de Paris, among others. He was made a fellow of the Linnean Society and of the Royal Society. He was awarded two honorary doctorates, one from the University of Dublin, the other from Oxford.

He was a name famous enough in the science of evolution to be invited to go on a lecture tour of the United States. In ten months there he covered a wide range of territory. In West Virginia he caught up with William Edwards, whose book nearly forty years before had pointed Wallace and Henry Bates toward the Amazon. Edwards had a pleasant house in a pretty orchard at the foot of a mountain and a fine collection of North American butterflies. Four days with Edwards and his specimens was not too long. A short conversation with President Grover Cleveland was more than enough; Wallace was not impressed. Overall, he came away from the United States impressed and depressed in about equal measure. He classified Americans as being in the front rank for inventiveness, deep devotion to science, and love of nature. But at the same time, in a mad race for money they were wrecking their country, chewing up the land, destroying forests, exhausting natural resources of oil and gas. And the big American cities were crowded and reeking, humanly awful; wealth monopolized for the aggrandizement of the few was leading to poverty, degradation, and misery for the many.

Wallace's American expedition was his longest after the Malay Archipelago. Into the beginnings of his old age, he and his wife went on walking tours around England, to Wales, to the top of Mount Snowdon, and once in a while to the Continent, sightseeing and botanizing. He never went back to the tropics.

Decade by decade, though, the tropics came back to him. In 1874 a letter was delivered all the way from Aru. It was dictated by an islander who could not write, to a scribe who took it down in Arabic. In English translation, it said that the islander was the man who had guided Wallace along the waters of Wanumbai. He still had a silver-headed cane that Wallace had given him, and he held Wallace in such high regard that he thought he should come out and govern Aru.

At home in the English countryside, Wallace had hundreds of subtropical plants growing. People kept sending him orchids from all over the world, and every so often one would arrive bringing with it the chrysalis of a moth or an interesting live beetle—even a new species creeping out to be classified and named for science.

In 1907 another letter came from the Malay Archipelago, from Ternate in the Moluccas, where in 1858 Wallace's paper on evolution by natural selection had been mailed to Darwin. This letter was from an American, a Harvard zoologist in the field, writing to say that he had met a wizened old Malay man in a faded blue fez who introduced himself as Ali Wallace. He enclosed a photograph he had taken.

Forty-five years earlier, when Wallace was leaving the islands, he had had a photograph made of Ali, dressed in Western clothes; and he carried the picture home with

Ali, Wallace's invaluable Malay companion and assistant, photographed in Western dress just before Wallace's return to England in 1862. (Wallace 1905, vol. 1)

him. Now here was Ali in old age still identifying himself with Wallace. Wallace, in his own old age, wrote to the zoologist, delighted and moved to recall the close association. Ali had done everything Wallace asked of him, and more. He was quick to learn how to shoot and skin, and he helped others pick up the skills. He helped Wallace learn Malay. He was a skillful boatman, calm even in storms at sea. He was trustworthy. He looked after Wallace during a terrible attack of malaria, in fact saved his life. As a parting gift, Wallace gave Ali two double-barreled guns, tools, stores, sundries, and money, enough to make him a rich man among his own people.

At home Wallace suffered from spells of recurrent malaria, reminiscences of the ague with changes in the weather, and volcanic eruptions of tropical-looking boils—the jungle coming back to him, or back at him, in his body. For all that, he never held the common nineteenth-century European view that life in the tropics was bad for the white man. There was a whole imperial-colonial theory and practice of what was called "tropical degeneration": the wearing of a special hat, called a solar topi, to protect against the sun, and a belt to ward off cholera; the necessity of regular retreat to hill stations and frequent home leaves; the danger of inertia and neurasthenia. Wallace did not subscribe to any of this. Between ages twenty-five and thirty-nine he had lived for twelve years in two different tropical regions. Not only did he survive, he lived another half-century and more, and he was ready to say that he owed his long life positively to his tropical years. In his view, all that was needed for good health in the tropics was work. If the white man made the native do all the hard labor, it turned the native into a slave and the white man into an illness-prone loafer.

A portrait of Alfred Russel Wallace before his death at ninety years of age. (Society for Psychical Research / Mary Evans Picture Library)

Wallace lived and worked on through the end of the nineteenth century and into the twentieth. In 1908, his eighty-sixth year, he was honored in two ways. He received the Order of Merit, Britain's highest award for a long life of distinction in science, literature, or the arts. And on the fiftieth anniversary of the historic joint Linnean Society presentation of 1858, the society created a Darwin-Wallace medal and invited Wallace to be the first recipient.

Darwin had died in 1882. At his funeral Wallace was one of ten pallbearers, bringing up the rear of the coffin. Before and after Darwin's death, Wallace always spoke of him in the highest and most generous terms. In 1889, when Wallace published his own book on evolution by natural selection, he titled it *Darwinism*, not Wallaceism. Now, in his speech at the medal ceremony in 1908, he gave all honor and priority to Darwin for the great discovery that he, Wallace, had made independently.

Wallace saw himself and Darwin as having arrived at the discovery separately but by the same path of life experience. Both had the collecting passion, an intense, almost childlike interest in the variety of living things. Both when young were ardent beetle hunters, and both seized on the fact that beetles had an infinite number of forms and innumerable adaptations to diverse environments. Both came across Malthus and his powerful theory of the struggle for survival. Both were travelers, which forced upon them an awareness of the strange phenomena of local and geographical distribution. And each during his travels had solitude, ample time for reflecting on daily observations. Collecting, travel, wide reading, deep thought, solitude—this was the Wallace formula for a life of original, productive work, and it was true enough of Darwin too.

For his Linnean Society speech, Wallace made one of his rare trips to London. At home he liked to work outdoors in his garden and his greenhouse. When he grew too old for gardening, he had his favorite plants arranged so that he could look at them through his study window. Inside in the study, he was always at work. He published a book at ninety and had two more planned. On November 7, 1913, just two months short of his ninety-first birthday, and close to sixty years after he sailed for the Malay Archipelago, he died in his sleep.

Wallace in the archipelago had been Western Man, Civilized Man, coming to terms with the philosophical truth that the living things of the forest were not organized with exclusive reference to the use and convenience of man. His first action once back in England in 1862 provoked repercussions for the living things of the forest all the way to the end of his life in 1913: he brought living birds of paradise to a society and a culture in which it was taken for granted that all living things were for the use of man.

And woman. Bird of paradise plumes looked wonderfully decorative on women's hats. Before Wallace's adult years, the plumes had been high fashion in the capital cities of Europe, London included. The vogue had subsided. But fashions could come back. This one did, and if any single thing set it off again it was the public sensation of Wallace's two male birds with their showy plumage, drawing crowds at the Zoological Society gardens.

While Wallace in his earliest days back at home was still preparing a paper to be read to a scientific audience on his naturalist's search for birds of paradise, the *Times* of London was making a fashion prediction: "We shall be, indeed, astonished if Paradise-birds' plumes, which are stated by Mr. Wallace to have been for several years declining in price as an article of trade, do not again rise in the market, as these living producers of them

Bird of paradise plumes were a popular fashion accessory for women through much of Wallace's lifetime. (Mary Evans Picture Library)

become more generally known, and recover their now almost forgotten value as ornaments for the hats of our fair countrywomen."

This happened, and on a scale that Wallace would never have predicted. The Law of Unintended Consequences had not yet been formulated, but here was an outstanding example.

The plumes of the bird of paradise were not the only ones coveted for fashion in the second half of the nineteenth century. Feathers and more feathers were the rage. All kinds of birds from all over the world were hunted: kingfishers, parrots, jays, magpies, goura pigeons, plotus, swans, marabous, tanagers, orioles, ibis, emus, rheas, turkeys, vultures. Owls' heads and wings turned up on hats. So did whole sea birds, and stuffed hummingbirds on a branch.

The feather trade was huge. Between 1870 and 1920, something like twenty thousand tons of ornamental plumes came to Britain. A single British trader said that in a single year he sold the plumage of two million birds. There were warehouses in London where, so a magazine journalist wrote, "it is possible for a person to walk ankle-deep—literally to wade—in bright-plumaged bird-skins, and see them piled shoulder-high on either side." Thousands of bushels, the *Times* said; enough parrot skins alone to cover much of Trafalgar Square.

These British figures did not include ostrich feathers from South Africa. And London was only one market; imports into France over a parallel stretch of time were more than double the British. Millions of birds were killed worldwide for fashion every year.

In the big markets of London, Paris, Berlin, and Amsterdam, the leading attractions by volume were plumes from ostriches, egrets, and pheasants. Demand was so great that all three were bred in captivity for killing. By price, the most desired item was the plumage of the bird of paradise. It was hunted in the wild.

Even before Wallace's time in the Malay Archipelago, Dutch bird of paradise traders were sailing from Ternate to New Guinea to hunt and collect. Rennesse van Duivenboden made a good thing out of this, from the 1830s on. His son-in-law went into the trade too, and there were others. Van Duivenboden's principal market at first was not fashion but science. He sold to merchants in Europe who sold to museums and to patrons of scientific institutions, at premium prices for new or unusual species. When the fashion craze took off, the market grew so much that by the 1880s dozens of plume hunters were operating out of Ternate. And by the turn of the twentieth century there were hunters living in New Guinea—colonial settlers, copra planters, Dutch, German, British, and Australian. Cash from plumes was often what kept them going through lean times until their first crop came in. In those small struggling colonial economies, bird of paradise

The flamboyant colors of a male Lawe's Parotia, a fruit-eating bird of paradise that inhabits the lower montane forests of Papua New Guinea. (Bruce Beehler)

plumes were a significant item of trade. Especially in Dutch and German New Guinea, there were years when plumes were the biggest earner of export income.

The bird of paradise plume boom in its peak period was like a gold rush. At the London auctions between 1904 and 1908, something like 150,000 skins were sold, an average of 30,000 a year—which meant that birds of paradise were being killed at the rate of one every twenty minutes, day and night.

The boom lasted until 1913; World War I effectively brought it to an end. Wallace lived through it all, from the day he brought his two birds of paradise alive to London, to a time when there was genuine alarm that the most fashionable species, meaning the showiest—the greater, the lesser, the red, and the Raggiana—might be hunted to extinction.

Over the course of this long, long boom, two generations of naturalists followed Wallace to the islands of the Malay Archipelago and New Guinea. They still had as their primary scientific aim and impulse the collection and preservation of specimens, meaning the killing of birds and animals.

By the last years of the boom, scientific interest was beginning to broaden and deepen, to include the preservation of living species. As one expression of this new impulse, some birds of paradise were taken from New Guinea to the West Indies, where a breeding colony was established on the island of Little Tobago.

Worry over the future of birds of paradise was part of a broader concern about bird extinctions generally. The reverse side of the plume boom was the rise of conservationist sentiment in the late nineteenth and early twentieth centuries, including the formation of bird protection societies and the development of a range of arguments against mass slaughter for fashion. This in turn led to sustained campaigns on both sides of the Atlantic to halt or at least limit the plume trade for all species. In the New Guinea colonies, the Dutch and the Germans brought in licenses and hunting seasons for the bird of paradise. In Papua, the Australians outlawed the taking of plumes altogether. And by the middle 1920s there were laws on the books of both the United States and Great Britain making trafficking in plumes a crime.

At the end of the twentieth century, there is no scientific need to collect more bird of paradise skins; the Museum of Natural History in New York, for example, has between three and four thousand skins, stored for decades and never studied. And these days, much bird of paradise science can be and is being done in the laboratory—DNA testing for evolutionary relationships, electron scanning microscopy of wire plumes, the aerodynamics of tail feathers.

Just the same, many questions can still only be addressed in the field. For a given species, is mating behavior monogamous or polygynous? What does this have to do with male plumage and display? What is the relationship between bird and forest? What is the range of movement of the bird within the forest? Does this vary with year-by-year changes in climate? What determines range—food supply from fruit-bearing trees? How many different kinds of trees? If you took away those trees, what would happen to the fruit-eating bird? Or turn the question around: Take away the fruit-eating bird, and hence remove a disperser of seeds, and what would happen to the forest?

Those are ecological questions. And of course the same sorts of questions need to be asked and answered not just for birds of paradise but for all species of birds, and not only birds but animals and plants as well.

In this kind of scientific inquiry a clear line of descent is traceable: back to Wallace, to his zoogeographical observations, which were fundamental to the establishment of the modern science of biogeography in the 1860s and 1870s; and then forward to the realization a hundred years later that, within biogeography, it is islands that offer the most striking theoretical insights and present the clearest practical lessons—insights and lessons that are the basis of the recent and increasingly urgent world interest in biodiversity and conservation.

In fact, the importance of islands in conservation thinking goes back to before Wallace and ranges forward to the present and into the future. From quite early in the age of European exploration and colonization, islands were seen as the world's prime test cases for what civilized Western man was doing to alter natural environments, all the way from transforming local landscapes to changing the climate of the planet by the large-scale removal of forest cover. Those were questions taken seriously in the seventeenth, eighteenth, and nineteenth centuries. They are even more serious questions for the twentieth century, and the twenty-first. And nowhere in the world are they more important than in Indonesia.

8

Floating market in Kalimantan. Many people depend on the cornucopia of foods and products provided by Indonesia's forests and seas to meet daily needs. (Greg MacGillivray Films)

FROM THE MALAY ARCHIPELAGO TO INDONESIA 8

Spanning the Centuries

Late twentieth-century Indonesia is much better mapped than the Malay Archipelago of Wallace's time, and the biogeographical inventory of the region is far more detailed. There are, in round figures, 17,500 Indonesian islands—making up 1.3 percent of the total land area of the earth—with more than ten thousand species of trees, about a tenth of the world's flowering plant species, about an eighth of all mammal species, nearly a sixth of all reptile and amphibian species, a sixth of all bird species, and about a third of all fish species. In almost all plant and animal taxa, Indonesia has levels of species diversity and endemism that rank with the highest in the world. And among humans, Indonesia has the world's greatest ethnic, linguistic, and cultural diversity.

It also has the world's fourth largest population: 206 million and rising. By sheer weight of human numbers, nature has been put under severe pressure. Indonesia already has the world's highest numbers of threatened and endangered plant and animal species. And given projections for human increase, the situation for nature is not likely to improve. Over the last quarter-century, life expectancy in Indonesia has risen by 50 percent; and at present rates of growth, within five decades the population will be more than 300 million—in those islands, an unprecedented rate of change among humans, with unprecedented consequences for nature.

Long, long ago in the archipelago, extinctions occurred, but only at the pace of geological and geographical change. The fossil record tells the story. On Sulawesi, for example—Wallace's Celebes, his island of anomalies—more than one million years ago there used to be a huge pig, an elephantid, and a tortoise with a shell six and a half feet across, wider than Wallace was tall. They were gone before humankind arrived. With human settlement of the islands and then the clearing of land for the systematic

Opposite: Indonesia covers only 1.3 percent of the earth's surface, yet it harbors a disproportionately rich treasury of biological species. A wide range of habitats, including natural rain forests like this one in East Kalimantan, protect countless endemic species. (Jez O'Hare)

The garish bloom of the *Rafflesia arnoldi*—at almost a meter in diameter, the largest flower in the world—attracts fly and beetle pollinators in the forests of Sumatra and Borneo with its scent of putrefying meat. With no leaves, roots, or stems, it parasitizes the *Tetrastigma* vine for nourishment. (Alain Compost / PhotoBank)

Right: The black orchid of Kalimantan, one of the thousands of orchid species found in Indonesia. (Alain Compost)

Jakarta, a city of over ten million people, is the business and government capital of Indonesia. (Chris Stowers / Panos Pictures)

Left: An aerial view of a crowded urban area of Jakarta. The population of Indonesia has already surpassed 206 million. (Mathias Oppersdorff / Photo Researchers, Inc.)

Airborne pollution chokes Jakarta; other forms of pollution—chemicals, sediments, agricultural fertilizers, and pesticides—affect riverways and coastal areas. (Jez O'Hare)

Monoculture plantations for rubber and oil palm have replaced species-rich lowland rain forest in many places in Borneo and Sumatra. Here a huge area has been cleared to establish a new plantation. (Frans Lanting / Minden Pictures)

cultivation of rice, natural habitat began to shrink, especially for big animals like the elephant, the rhinoceros, and the tiger. When spices began to be cultivated on a large scale for the markets of Asia and later Europe, more land was cleared, more habitat lost. In the nineteenth century, plantation agriculture—coffee and tobacco—meant yet more habitat loss. And in the twentieth century, especially with the population explosion of recent decades, the process has continued ever more extensively and ever more rapidly.

Severe pressure on nature is not simply a domestic matter of making provision for a rapidly growing number of Indonesians at rising increments of human needs and expectations. The Indonesian economy is part of the world economy, and in the global market Indonesia makes money out of its natural resources—oil, minerals, and the rich productivity of volcanic soil, coral reefs, and rain forests. In recent times, small-scale agriculture has increasingly been supplanted by large-scale plantation monoculture: cash crops of rice for domestic consumption and coffee, rubber, and palm oil for export. In the process, riverine systems have become polluted. Coastal fisheries have been damaged, reefs destroyed. And forest, including rain forest, has been cut down and the cleared land burned to allow agriculture and grazing.

Indonesian hardwoods are highly desirable and readily marketable overseas, and it is in the logging of timber that the rapidity and extent of the transformation of nature are visible in the starkest outline. As recently as three decades ago—in some areas even two decades ago—it would still have been possible to see a Malay Archipelago forest landscape much as Wallace saw it. Now, in many places, whole Indonesian forests and their associated plants and animals have simply disappeared.

On big islands, logging starts in lowland valleys and advances upward. Spreading to smaller islands, it has become a dominating fact of economic life and a daunting factor in natural life even in tiny out-of-the-way places—all the way to the Aru Islands, where Wallace saw his first birds of paradise in the trees. The Aru hunters had an experienced, intimate knowledge of their forest, down to individual display trees, just as they knew down to a single bird how many fully plumed males they could take in a given season without permanently depleting the stock. With the Aru forest itself depleted by logging, the bird of paradise population is no longer in equilibrium. And what is true for the bird of paradise of Aru is true Indonesia-wide for any number of other species, from the tiger to the hornbill to the orangutan.

The logging companies are wide-ranging and omnivorous. They can find a market for more than a hundred species of trees, to be milled into timber or pulped to make paper. As one biological scientist puts it, the forest with the highest incidence of endemic species can also be the forest with the highest incidence of logging companies.

The economics of Indonesian timber and other exports are not isolated in their impact on nature. They intersect, collide, and tangle with the domestic politics of Indonesian population growth.

Of the approximately 17,500 Indonesian islands, about 6,000 are inhabited. But between 60 and 65 percent of the total population lives on just three islands, which together make up only 7 percent of Indonesia's total land area: Bali, Madura, and Java (Java, with 800 people per square kilometer, is one of the most densely populated places on earth).

Indonesian government policy is to encourage relocation—called transmigration—out of these heavily populated areas, to Sumatra, from there eastward to Kalimantan and Sulawesi, and finally to the largest province, Irian Jaya.

Wallace knew Irian Jaya as the western part of New Guinea. For him it was *terra incognita.* It has the highest mountains in the archipelago and the most rugged geological formations; it is where two of the earth's tectonic plates grind together. Until recently its population was perhaps a million indigenous tribal people, who lived by hunting and gathering or subsistence farming. Now something like a quarter of a million Indonesians

Fueled by the world demand for tropical hardwoods, Indonesia's rapidly expanding logging operations are decimating the lowland rain forests of Sumatra and Borneo. Approximately one million hectares of forest are cut each year, and logging roads open vast areas of the interior of Borneo to small-scale agriculture, further eroding the integrity of rain forest ecosystems. (Frans Lanting / Minden Pictures; PhotoBank)

Transmigration settlements in Indonesia's relatively unpopulated outer islands promote agricultural schemes that are often short-lived because of nutrient-poor soils. (Jez O'Hare; Charly Flyn / Panos Pictures)

Copper and gold mine in Irian Jaya, the largest deposit in the world. (Kal Muller / PhotoBank)

have settled there, most of them as transmigrants. Oil has been found in the lowlands. In the mountains, one of the world's biggest mineral companies is exploiting the world's biggest deposit of copper and gold. The province is mostly rain forest, and rain forest is a leading source of hardwood. These days Irian Jaya, like the rest of Indonesia, is being traded to the world.

Just as Indonesian hardwoods are a lucrative source of income on the global market, so are Indonesian reef fish. On a world map of coral reef diversity, the rich center is a triangle taking in the Great Barrier Reef of Australia and the Philippines, then extending west into Indonesia, to Sulawesi and Kalimantan. In terms of reef fish, Indonesian waters have something like 2,500 species, about 35 percent of the world total.

Indonesian seas harbor the richest diversity of coral and fish species in the world. (James Watt / Animals, Animals)

The green sea turtle was once commonly encountered in Indonesian seas. Today its existence is increasingly threatened by humans who prey upon it for meat, oils for cosmetics and medicine, and shells for jewelry. (Kal Muller / PhotoBank)

The world has a double appetite for these fish. The international aquarium business wants showy, high-priced Indonesian species to sell—triggerfish, angelfish. The annual trade in live Indonesian reef fish for aquariums is 25,000 tons. At the same time, the Asian restaurant business wants high-status, high-priced exotic seafood dishes on the menu: lips of the Indonesian humphead wrasse, $225 a plate.

The most efficient modern way to catch reef fish is to have skin divers go down with squirt bottles of poison—sodium cyanide. The cyanide stuns the fish. They are kept penned in water until the toxin works its way out of their system and they can be shipped away for sale.

For every one high-priced fish that is harvested, hundreds of other fish die. All kinds of invertebrate reef species too, and the coral polyps that they feed on.

Cyanide fishing was developed by East Asians, and Indonesian fishermen have learned to use it, though it is illegal there. Middlemen supply them with the cyanide, which is manufactured in the United States and China, and then broker the catch to overseas buyers. And increasingly, foreign fleets of Asian fishing boats are in Indonesian waters, using cyanide.

When reefs are fished out beyond natural replenishment, the cyanide fleets go looking for somewhere new. Yesterday it was the Philippines, where 80 percent of fishing grounds have now been overharvested and essentially abandoned as unprofitable. Today it is Indonesia. By the end of the century—meaning really only tomorrow—the international reef fish business in Indonesian waters will have ceased to be profitable on the world market, and what Indonesia will be left with is dead, poisoned coral reefs.

It is not just cyanide that wipes out fish populations and destroys reefs. There are other destructive practices, including fishing by dynamite. Even where the catching method is not poisonous or explosive, overfishing is increasingly common. And Indonesia-wide, industrialization on land excretes pollution that washes damagingly into the sea.

A survey and assessment of the nation's reefs in the late 1980s and early 1990s found that in western Indonesia, the most highly developed area, more than half the reefs were in very bad condition. In central Indonesia the figure was more than a third, and in eastern Indonesia the situation was only a percentage point or so less serious. Remoteness from big population centers is no guarantee of exemption from damage to marine life. In the waters of Teluk Cenderawasih, the shells of giant clams are empty, the meat scooped out for the export market. The harbor of Ambon, with all its amazing richness of marine life that was such a revelation to Wallace, is dead and dirty.

The sum of modern biological knowledge about Indonesia—what is understood in Western scientific terms about nature, its past history, its present condition, its future problems and prospects—has been put together by successive generations of naturalists going into the field like Wallace.

As of the fourth quarter of the twentieth century—meaning the fifth and sixth generations of researchers after Wallace—technical advances have been made in equipment: drug-filled dart guns instead of shotguns, geopositioning devices instead of compasses, notebook computers instead of pen and paper. Living conditions have improved considerably: plastic sheeting for instant shelter, solar-powered lights, thermos flasks for ice water, antibiotics. And transport is better and faster: airplanes for the major long hauls; boats with engines between islands, around the coast, and upriver; helicopter dropoff and pickup in the mountains.

Not that in Wallace territory motorized research is any guarantee of speedy results. Sulawesi has a carnivorous civet cat, *Macrogalidia musschenbroekii.* Wallace did not collect it. In fact, it was not known to Westerners until late in the nineteenth century. After that, specimens were acquired by some European museums, but as late as the mid-1970s it had never been scientifically sighted and photographed alive in its native habitat. To catch it on film took three tries over three years—each attempt involving a flight from Jakarta, a day's drive into the interior, the negotiation of permissions from authorities, meetings with local headmen, and tactical discussions with village hunters and rat catchers concerning possible civet cat whereabouts, followed by days of scrambling along ridges and stumbling through valleys. In the first two annual searches, all that was sighted of *M. musschenbroekii* was droppings. Not till the third year was the animal's picture caught in a camera trap.

One difference since the time of Wallace on Celebes was that a strange-looking white man doing strange things no longer provoked terror and flight in the villages; the scientific civet tracker was himself tracked by hordes of children.

Wallace in his day was professionally alone. Over the next generation or two, the number of naturalists in the field in any given year would have been no more than a handful. Today, the number is in the hundreds. As one scientist says about his years of experience in Indonesia, when he started he knew everyone in the field of conservation biology; now he does not.

Numbers are up, then, and as well there are late-twentieth-century changes in the composition of the profession. The present generation of fieldworkers includes Indonesians with university degrees, not just the "native helpers" of Wallace's day. And the modern Indonesian professional contingent includes women, which in Wallace's time would have been unimaginable.

One of these women, a marine biologist, has been out on the waters of the eastern archipelago like Wallace. She has navigated the Wanumbai Channel in the Aru Islands, has seen the forests reflected in the smooth waters, the sun going down in an orange glow. She has also been seasick, bitten to distraction by insects, shipwrecked on a coral cay, caught in a hurricane, and surprised ashore by a small earthquake-generated tidal wave that came out of nowhere and temporarily buried one of her co-workers in the sand.

From Wallace's time to now, hardship in the field for the sake of science has been a matter of course: it goes with the territory. As a near-certainty, researchers could expect to experience more leeches than letters from home. With bad luck, they might be hooked bodily out of a dugout canoe by ferocious spikes of rattan overhanging a stream. With worse luck, they would come down with tropical sores that refused to heal, or with malaria, modern drugs or no. On the side of good fortune, they would have knowledgeable villagers for carriers, guides, and companions. And for teachers. Someone in the forest knowing how to survive without a gun or a knife, how to treat a pit viper bite. Someone on the coast knowing the lunar calendar for the spawning of the grouper. Inland again, someone knowing all the varieties of durian and all the forest paths to the best trees. Someone else knowing every kind of rattan—where in the trackless jungle to find a single plant of a particular variety, and when to find it in full fruit: "I met it several years ago, and I thought it would be fruiting now."

Wallace was by no means against the idea that island people possessed useful factual knowledge and had systematic ways of understanding natural processes. He took information where he found it and was appreciative. But he was a white man of the nineteenth century, and it simply would not have occurred to him to name a new species after the islander who first brought a specimen to him. The number of Malay Archipelago species bearing Wallace's name is in the dozens. The number of species bearing the name of his invaluable companion and forward scout Ali is zero.

Toward the end of the twentieth century, the names of Indonesians are appearing in formal scientific identifications. Call this cultural evolution.

From Wallace's time to the present, the naturalist has experienced the engaging oddity of being simultaneously observer and observed. For island people, Wallace the collector was an entertainment, a show. A hundred years later and more, a twentieth-century rattan botanist, deep in the forest, alone, absorbed in his work, gets the feeling that he is under surveillance. He looks up, directly into the gaze of a siamang gibbon in a tree contemplating this strange evolutionary relative playing with plants on the ground. The siamang watches the scientist for five minutes; the scientist watches it watching him, then watches as it brachiates away to join its troop howling in the distance.

The personal rewards of professional fieldwork today remain the same as they were for Wallace: the discovery of a new species, or a changed perception of the patterns and processes of natural life, and the sense of fulfillment that comes with such understandings.

In the great dipterocarp forests of Borneo—forests that Wallace experienced as illimitable and which are now being disappeared—a late-twentieth-century scientist asks a question: How are the flowers of such towering trees pollinated? Whatever the process, it takes place 150 feet up and higher. How to find out? Rig booms in the tree canopy like artificial branches, then sit and watch. The answer that comes down with the scientist from the majestic heights is: thrips. They lay their eggs in the buds. When the flowers are in full bloom, the corollas fall in drifts to the ground each morning. Each night thrips emerge, covered with pollen, minute winged insects so small that to them the air is as thick as water, beating upstream like tiny oarsmen, up to other flowers in other trees, pollinating.

For the scientist, experiencing the flowering of the dipterocarps is close to sacramental—the forest as a great gothic cathedral of nature, two- and three-hundred-foot trees buttressed at the base, slender-trunked, the branches high above fanning out into a lofty vault, the light through the leaf canopy emerald green with golden sunflecks, the air filled

Towering dipterocarp trees can reach heights of 75 meters, providing the structural framework for species-rich lowland rain forest ecosystems in Borneo and Sumatra. Unsustainable logging practices for these commercially valuable species are resulting in irreversible degradation of the forests. (Jez O'Hare)

Winged fruit of the dipterocarp tree. The synchronized mast fruiting of many dipterocarp species occurs every two to seven years, providing a bonanza of oil-rich seeds for forest animals. (PhotoBank)

with the echoing sound of birds and insects. The huge trees flower only once in four years or even less often, sometimes only once in ten years. Then the air is heady with perfume, petals falling like snowflakes, piling into fragrant drifts underfoot. A rare experience, arousing emotional intensity comparable to Wallace's collector's ecstasy. And as more and more of the dipterocarp forests are logged, clear-cut out of existence, it is an experience rarer and rarer.

Loss of forest means loss of habitat, which means endangerment of species. The orangutan is an extreme example of an animal mercilessly squeezed by modernity. The fossil record shows ancestral orangutan bones all across Asia, from Peking to the Malay Archipelago. The orangutan of Wallace's time was already reduced in its range to parts of only two islands, Borneo and Sumatra. Among nineteenth-century men who hunted freely, as Wallace did, it would still have been possible to believe that however many orangutans they killed, there would still be more. Not so any longer. Orangutans are an endangered species—the most highly endangered of the great apes—and have been recognized as such for at least a quarter of a century. Their numbers are down radically, and only 2 percent of their habitat is protected.

Young orphaned orangutans at a rehabilitation center in Sepilok Reserve, Sabah, being transported to an exercise area in the forest where they are taught to climb trees. Infant orangutans are captured from the forest for the pet trade but are often abandoned when they mature. Many of these animals have been brought to rehabilitation centers in Sumatra and Borneo, which attempt to reintroduce them to life in the forest—with limited success. (Frans Lanting / Minden Pictures; Michael Nichols)

How many orangutans can survive in the fragment of habitat that remains to them? Is it possible to help orangutan populations to sustain themselves? And—a very late-twentieth-century question—is it possible to reintroduce captive orangutans to the forest, to rehabilitate them from civilization to the wild?

Why should a question like that last one even have to be asked? In scientific terms rehabilitation is a serious problem, and it has been exacerbated by the whims of fashion.

Orangutans have always been in demand as pets. In modern times, the demand has been met by shooting adult females and smuggling their young away for sale. At the start of the 1990s, a craze for pet orangutans was set off in Taiwan when one made an appearance on television. By 1993, more orangutans had been sold illegally to Taiwanese than there were in all the zoos in the world.

In the fallout of the Taiwan case, smugglers were prosecuted and smuggled orangutans were confiscated. Some of the animals were returned to Indonesia.

They were not the first candidates for orangutan rehabilitation. Experiments had been going on for twenty years, with serious concerns faced all along the way. What if remaining habitats were already close to the limit of carrying capacity? What were the pros and cons of introducing strange animals into existing populations? Would it be better to look for other areas of possible habitat currently without orangutans? But then, what about introducing animals—large animals at that—of a species entirely new to a particular area? And wherever orangutans might be put back into the forest, what if animals from civilization were carriers of diseases of civilization?

Questions persist, answers are still being explored. Meanwhile, the number of orangutans in Borneo stands at about 15,000. In Sumatra it is about 5,000. Computer models indicate that if an orangutan population falls below 5,000, extinction will follow.

There are perhaps only a dozen scientists studying the life and critical times of the orangutan. By Indonesian standards, this is a heavy concentration. Nationwide, conservation biologists, even in the increased numbers of recent years, are far rarer and more thinly spread than orangutans, averaging out at fewer than one per major conservation problem.

Morowali, a nature reserve of about 225,000 hectares in central Sulawesi, is typical of Indonesia's conservation areas in that it poses a whole range of interconnected human and natural questions. The reserve runs from mangrove swamps and coastal lowlands to steep mountains inland, rain forest and sub-alpine forest, and five strongly flowing rivers. Among designated parks, Morowali is unique in Indonesia, perhaps in the world, in that it has the largest swath of forest with ecosystems finely adapted to ultrabasic soils that are very nutrient-poor.

Morowali Nature Reserve in Central Sulawesi harbors the largest tracts of unusual forest adapted to nutrient-poor ultrabasic soils. (Jez O'Hare)

Rattan collectors are everywhere, their tracks crisscrossing the traditional trails, cutting deep and deeper into the forest as they overharvest, leaving big host trees felled and habitat disturbed.

And Morowali has other problems generated by the clash between tradition and modernity. The park is the home of the Wana. They are a mountain people, a rain forest people. Their way of life has always been one of moving from place to place, hunting animals and growing corn, tapioca, rice, and tobacco in shifting gardens cleared successively from the forest. They never wanted much to do with civilization. They avoided the Dutch colonial government, and until recently they had little to do with the Indonesian government. They kept—conserved, so to speak—their own language, religion, culture. Now their traditional home is a nature reserve. But are they part of nature?

Under national park regulations, human habitation in parklands is not approved. Official policy is relocation to settlement camps outside park boundaries. But is that the only option? Might it be possible to reconcile the needs of the nature reserve with the life of the Wana?

Before such a question can even be addressed, baseline information is needed. How many Wana are there? How many animals are there? Which ones do the Wana hunt? How many do they catch? And how much forest do they turn into garden?

These are scientific questions. How to find answers? Have scientists live among the Wana.

This is arranged. Two of the scientists are biologists. They are Indonesians, university-educated in Indonesia, one of them a woman. The other scientist is an anthropologist; he is American.

He knows about Wallace; he has read *The Malay Archipelago.* And he has his own Wallacean stories of living among people who have their own view of the world and what is in it. A late-twentieth-century Wana of wide worldly experience, he says, is one who has made the bus ride to Palu, along a bad road, eighteen hours each way on a good day—when tires do not go flat and the brakes do not catch fire. Most Wana have not traveled as far as Palu. They live and move and have their being wholly in the forest. To them, all white people look alike, and ghosts are real. They have difficulty with the idea of the United States, a place where ice falls from the sky; six months to walk there and impossible anyway because there is water in between. The anthropologist, for his part, has a shortwave radio on which he can pick up Voice of America and hear the country music of the late great Patsy Cline or a feature about the Elvis Is Alive Museum—electronic ghosts from his own culture.

A

(a–b) The Wana have maintained relative isolation from much of the outside world, and most adults speak little or no Bahasa Indonesia, the country's lingua franca. The Wana have minimal interaction with the cash economy of the modern world; they still adhere to a strong traditional belief system and depend on the forest for all their daily needs. (c) Wana blowpipe quiver. (Michael Alvard)

B

C

The Wana have rules, to do with the dead, about where it is good and not good to build a dwelling. The anthropologist locates accordingly. He becomes used to living with no privacy, on an earth floor under a thatch roof, with rats, snakes, and centipedes, and cockroaches running around the rim of the cooking pot.

He sponsors a ceremony with shamans, featuring much eating, drinking of rice wine, and smoking. The Wana all smoke. The anthropologist develops a taste for Indonesian clove cigarettes and learns to chew betel nut.

In the rain forest his feet grow funguses. He sprouts the first boils of his life. They are volcanic, like Wallace's. The one on his left shoulder he names Mount St. Helens, the one near his right knee he names Krakatoa. He gets malaria. His temperature shoots up to 105 degrees; doses of quinine make his ears ring, and he hallucinates flying to quasars to witness the birth of the universe. Another fever may be dengue, or it may not. He has no doctor for diagnoses, only a health guide called *Where There Is No Doctor*. He supposes that in the tropics it is good practice to travel with a fever; it recalls all the great adventurers who passed his way. He knows how Wallace discovered the theory of evolution by natural selection in a fever. His wish is for his own fever to be as productive.

The forest where he works is all tangled slopes, roots, and rocks. The scientists' field station at Tambusisi, built for them by some Wana, sits on a terrace where a small stream joins the Morowali River, just above a big whirlpool. Trails cut uphill are steep; trails cut along the slope run into gorges hundreds of feet deep. There are mudslides. The anthropologist falls down a lot. He conceives a humorous theory of the origins of civilization, having to do with the development of smooth walking areas. Other than that, he has no time for great thoughts. At every step he has to look down so that he can plant his feet safely. But then branches and vines, many of them with barbs, smack him in the face. And what about the Wanas' concealed animal snares, pits with sharpened bamboo stakes, triplines with spears? A walk in this park is no walk in the park.

The Indonesian biologists walk measured transects in the forest, recording all encounters with birds and animals. The anthropologist collects bones, mandibles of pigs, anoas, and rats, and counts them. He is also taking a human census. There are not many Wana at Morowali, perhaps only five thousand; but accounting for them turns out to be an inexact science. Place of residence is hard to establish: people have more than one dwelling house and may have several garden houses. Names are harder still: adults are called by the name of their eldest child, and as people get older it becomes impolite to call them by their real name. Age is impossible: nobody knows dates of birth, their own or anyone else's.

As part of his study of the total impact of the Wana on the forest, the anthropologist is locating and mapping the sites of their dwellings, ceremonial houses, gardens, and trails. He has sophisticated handheld electronic equipment to check coordinates on the ground with a Global Positioning System satellite—latitude, longitude, and altitude. He can tell where the Wana do everything they do on the face of the earth, correct to a meter. This is a Western way of knowing.

The Wana know Morowali in a different way, because they have always lived there. Now the place where they have always lived has been designated a conservation reserve. What they are cutting down for their gardens is primary rain forest; what they are hunting with their spears, blowpipes, and dogs are rare and endangered species of animals unique to Sulawesi. The Wana are an indigenous people redefined by modern conservation thinking as a threat to nature.

Rapid population growth, transmigration, the expansion of large-scale monocrop agriculture, mining, logging for export—all are part of development in general, and thus integral to late-twentieth-century Indonesia, whose economy has been one of the fastest-growing in Asia. These activities all have their own powerful business and market dynamics, their own political drives and justifications, their own inherent social frictions and potential combustibility.

In mid-1997, two major Indonesian economic initiatives turned out to be incendiary—literally, and on a terrifying scale. On the large island of Sumatra, and on the even larger island of Borneo, productive land began to catch fire. The fires spread, burning for weeks and months, leaving hundreds of thousands of hectares scorched and blackened. The smoke thickened and drifted across Indonesia, district by district, region by region (fires broke out on Lombok and Bali, on both sides of Wallace's Line, and on Timor, where Wallace had picked up the wild honeycomb he took back to England and gave to Charles Darwin), then country by country across Southeast Asia: Indonesian smog blanketed Malaysian Sarawak (the part of Borneo where Wallace first saw an orangutan) and obscured the sun in Thailand, Singapore, Brunei, Papua New Guinea—all the way to the southern Philippines.

Among man-made environmental disasters in this or any other part of the world, these fires ranked with the very worst in terms of scope and impact on nature and humanity. Satellite images showed thirty Indonesian cities smogged in, and in the countryside "hot spots" of roaring flame numbered in the hundreds, at the peak more than a thousand. Animals were fleeing the burning forests. Birds were dropping dead out of the sky. Babies born in Sumatra developed lungs like those of ten-pack-a-day smokers

Fires have devastated vast areas throughout Indonesia in the last few decades, a result of catastrophic weather patterns that cause drought, combined with widespread unsound forest-clearing practices. (Jez O'Hare)

Smoke from the 1997 forest fires in Borneo obscured the sun throughout Indonesia, Singapore, and Malaysia. (Jez O'Hare)

before they ever saw sunlight. In Malaysia, air pollution in Sarawak hit world record levels, and Kuala Lumpur declared a health emergency. Singapore businessmen were going to their offices wearing surgical masks; looking out of the windows of their high-rise buildings over the city, they could not see as far as the harbor. In the Strait of Malacca, one of the world's busiest sea corridors, visibility was so bad it disrupted shipping and caused a collision that killed at least 38 people. Thousands of airline flights had to be delayed, diverted, or canceled; on Sumatra a passenger plane approaching Medan in heavy smog crashed, killing all 234 aboard. Mines, factories, and schools had to close. The immediate loss in productivity and revenue was staggering, across the board, from agriculture to industry to tourism. The long-term foreseeable negatives were dismal to contemplate, from irreversible loss of plant and animal species to widespread impairment of human health, caused by respiratory disease and famine.

And months into the disaster, the fires were still burning, consuming land and life. Tens of thousands of volunteers pitched in: bucket chains a mile long were formed in Borneo, firefighters were flown in from Malaysia and Australia, water bombers arrived from Australia and the United States. Nothing could put an end to the burning—nothing, that is, short of the monsoon.

The 1997 monsoon was late. And before that the year had been a dry one. This was a consequence of El Niño, the huge weather system originating in the southern Pacific that repeatedly causes unsettling perturbations in climate everywhere from North America to Australia to Asia. The accumulation of massive areas of heated ocean water off the coast of Peru means colder water away to the west in the Indo-Pacific, which means lower relative humidity, which means less rain.

El Niño comes every two to seven years. The 1997 El Niño was as severe as any in the past half century. For Southeast Asia, including Indonesia, this meant drought, and drought meant high fire danger.

Indonesia in 1997, then, was a fire disaster waiting to happen. Fire was built into the country's way of life and way of doing business, as an element of traditional small-scale tribal slash-and-burn agriculture and, far more formidably—by many orders of magnitude—as an instrument of modern large-scale monoculture, all peaking in the 1990s.

The 1997 fires were essentially set and kindled by two large-scale national economic initiatives, one domestic, the other connected with the world market. The domestic intention was to make Indonesia self-sufficient in rice production. The world market aim was to double Indonesia's production of palm oil for export by the year 2000.

In the 1990s, forest was being razed and planted with oil palms at the rate of a million hectares a year. After the land was logged, the quickest and cheapest way to clear it for plantations was to burn.

As to self-sufficiency in rice, transmigrant labor was being used in Borneo to drain peat swamps and convert them to rice paddies. The peat had been laid down over thousands of years, forming a layer that might be up to 20 meters deep. Peat in its natural swampy state is a sponge, absorbing moisture in the rainy monsoon season, releasing it slowly in the drier months. Dried-out peat is flammable. It can catch fire at the surface and go on burning deeper and deeper, spreading from the peat layer to underground coal seams. Those deep fires are almost impossible to extinguish, and hectare for hectare they put worse smoke into the sky even than forest fires. In Borneo there was peat that had been burning since the last severe El Niño in 1983; and more—far more—caught fire in 1997, set alight when forest fires went out of control.

Some forest fires were spontaneous, accidents of nature in a drought year. Some were small tribal-agricultural slash-and-burn fires, of the kind that had been used for hundreds of years by shifting cultivators. The vast majority, the biggest and most destructive by far, were the product of the rapidly accelerating forest clearing of the 1990s for plantation monoculture. More than 2 million hectares of Indonesia's remaining 100 million hectares of forest was destroyed in 1997.

Burning of that kind, on that scale, might have been cost-efficient—good business, narrowly defined. But in other ways it was bad business, and known to be so, outside as well as within Indonesia: neighboring countries had been complaining for years about drifting Indonesian smog. In 1994, large-scale forest burning was made illegal, but it continued because it was profitable. And then in 1997 everything went up in flames.

The satellite images showed fires burning on land licensed to as many as 176 companies, including some of the largest in Indonesia, with ties to the national government. In every sense the fires were big business, with political implications.

And as if all that fire and smoke was not catastrophe enough for Indonesia, the country was simultaneously embroiled in an Asia-wide financial crisis that broke in 1997—a disaster that, like the fires, seemingly came out of nowhere but had in fact been in the making for a number of years.

This financial crisis was the money equivalent of a regional, even hemispheric El Niño disaster. Country by country in Asia, bailouts in the scores of billions of dollars were being talked about as necessary. In Indonesia it was going to take $40 billion.

Obscured in the enormity of the Indonesian financial crash and its political and social reverberations—relegated from the headlines, likely to be neglected, even lost sight of—was the fact that the fires had created an environmental disaster in Indonesia on a world scale, comparable in its effects and implications to the burning of the Amazon rain forests.

The Amazon on fire, Indonesia on fire: the two places where Alfred Russel Wallace worked in the nineteenth century had been put to the torch in the late twentieth century, burning up in the biggest man-made fires on earth.

With help from the weather, the smoke from Indonesia's 1997 fires might have cleared after one monsoon season. But continued drought kept the fires burning into 1998, seriously in Sumatra, out of control in East Kalimantan.

The drought seriously damaged the rice harvest. The Indonesian government, which had targeted national self-sufficiency in rice as a top priority, was forced instead to import rice on a huge scale, four million tons in 1998. Yet even with this massive input, supply and demand were still not at parity, and distribution was inefficient.

At the same time, the working world of Indonesia was in wreckage. The financial meltdown had thrown millions into unemployment, while the rupiah was losing as much as 80 percent of its buying power, drastically raising the cost of the most basic necessities of life. Rice was up 300 percent in a year. In East Java a kilo of rice was costing a day of a farmhand's labor. As many as half the children under the age of two in that region were malnourished. Nationwide, as many as four out of ten Indonesians were living in poverty. The situation was bleak to the point of desperation.

The smoke pall over southern Sumatra, as shown in a photomosaic taken on 27 September 1997 from the Space Shuttle *Atlantis* over the Barisan Mountains (oriented from southeast to

northwest). The images were made by KidSat students using a remote-control electronic still camera. (NASA/Office of Earth Sciences, Johnson Space Center, STS086-215701 and 215637)

Food riots broke out, escalating to looting, burning, and death in the streets—an urgent expression of the undeniable need for thoroughgoing reform of the Indonesian economic system. This in turn would involve, as a paramount issue, the disentangling of the pervasive and tightly woven connections between big business and government—and the military, which exercised a controlling influence everywhere.

The ultimate political marker of the size and scope of the national crisis was reached in mid-1998, when, after more than thirty years in power, President Suharto, chief executive officer of the modernization of Indonesia as a nation and its entry into the global economy, stepped down, effectively forced from office. Strongman government was no longer strong enough to hold things together at the center. Indonesia had in fact never been a monolithic nation-state, and now radically diverse interests were finding irrepressible expression in tumultuous social disorder.

For the successor regime, innumerable problems loomed, each separately identifiable but many of them cross-linked, making solutions difficult even to conceptualize, much less enact. No one could see a clear path to a horizon of economic rehabilitation, much less set a time frame for necessary reform, across the board, from land tenure to regional autonomy to social justice. In a country where market considerations had long since supplanted tradition as the determinant of land use, where the rate of growth of the economy had been exalted above all else, and where prolonged boom had now gone suddenly and spectacularly and disastrously bust—where desperate poor people were tearing up golf courses to plant rice—what could be done to right things?

Looking at the issue only in money terms, Indonesia was going to be burdened with tens of billions of dollars of repayments for loans internationally organized and administered. Where was the money going to come from? The nation's wealth in modern terms was in its natural resources. Would these resources now suffer renewed and even accelerated exploitation?

Those were high-level questions. Down on the tortured ground of daily life where millions upon millions of Indonesians were now forced to subsist, minimum food and basic shelter were being threatened, either made too expensive by rampant inflation or simply not accessible or available anymore in fire-razed areas. The economic disaster turned into an ever more widespread environmental catastrophe: more and more forest being cleared for subsistence planting, more and more endangered wildlife being hunted, either eaten where caught or sold at market for cash to buy food for survival. Here were humans caught up in a species war to the death—Malthus in Indonesia.

The crisis was engulfing local communities, indigenous peoples and transmigrants, a generation of students at the cutting edge of reform thinking and activism, whole islands and regions, and the nation at large, as well as the global business and financial instrumentalities that were now so heavily involved in deciding Indonesia's future.

In a way, the severity of the crisis, the failure of so many systems and institutions, offered an unprecedented opportunity for reevaluation. Amid this welter of contested priorities, might a vision emerge that would offer another chance to protect natural resources and habitats of plants and animals, forests and reefs—a vision convertible to a set of practical applications for managing resources and for a socially equitable distribution of returns: in short, a vision of a society in which humankind and nature could depend on each other for nourishing sustenance?

For Westerners, the relationship between the big world and the tropics has always been a complicated one. The complications go very deep, all the way to fundamental European conceptions of nature, which have shifted and altered over the centuries, appearing, disappearing, reemerging: nature as paradise; nature as wilderness, to be tamed and dominated; nature as a resource to be mined, exploited. What might be the benefits of involvement in the tropics? Benefits for whom? And what might be the costs—costs to be borne where, and how, and by whom?

Expansiveness and changefulness associated with the workings of the big world can mean destruction of natural life and extinction of species in the tropics. At the end of the twentieth century, with all of Asia, including Indonesia, inextricably involved in the functionings and malfunctionings of a one-world economy, these are questions and responsibilities not just for Westerners but for Indonesians as well.

Wallace was sensitive to the deep problem of the relation of continents to islands while he was in the archipelago, and he remained so for the rest of his life. It surfaced most strongly in his thoughts and feelings by way of birds of paradise. He gave them the greatest prominence in *The Malay Archipelago*, devoting a whole chapter and more to them in the body of the text, and naming them on the title page along with the orangutan. As well, in some editions the two frontispiece illustrations were of the orangutan and the bird of paradise. Both of them were shown under attack. In both pictures it was natives doing the attacking—Dayaks surrounding an orangutan with spears and hatchets, Aru Islanders shooting at birds of paradise with bows and arrows. But when Wallace came to discuss the fate of birds of paradise, it was the attacking white man that exercised him.

The rarest of all beadwork in Indonesia is the *tampan maju,* or beaded tampan. This early piece features a vivid "tree of life," one of the oldest symbols found in Indonesian art, whose curved branches and ornate motifs evoke a centuries-long tradition of respect for nature. (Collection of Frank Wiggers / Bruce Carpenter, San Francisco)

Wallace, living in the loneliness of the long-distance collector, naturalist, philosopher, and Westerner, had to wrestle with tensions in himself, by himself. The struggle showed in the themes that were sounded when he put words on paper for *The Malay Archipelago* about his first sighting of the bird of paradise. Here he was at his most conflicted. He was talking to himself as much as to the world, and he spoke in many voices at once. He was a traveling collector with a living to make, and the long-barreled gun was his tool of trade. He was also a white man in the tropics—an unusual variety of white man, to be sure, but still in many ways a representative man of his era, a mid-nineteenth-century European, meaning a cultural chauvinist, confident of the superiority of Western civilization, cutting a swath through the world in a period of increasingly wide-ranging and assertive European expansionism. At the same time he was a pioneering natural scientist of great intellectual reach, capable of taking the long view of animal and plant life on earth. And beyond all that, at his best he was one of those human beings, uncommon in any century in any culture, with the philosophical capacity—and the moral willingness—to try to stand above the concerns not only of his own small human tribe but of his own species.

His own first sighting of a bird of paradise in its native habitat was also the moment of his first physical grasping of a specimen fresh killed by gunshot, warm in his hands but dead. That set off one of his deepest meditations, on the fate of creatures of the wild when civilization caught up with them. "I knew how few Europeans had ever beheld the perfect little organism I now gazed upon, and how very imperfectly it was still known in Europe. . . . The remote island in which I found myself situated, in an almost unvisited sea, far from the tracks of merchant fleets and navies; the wild, luxuriant tropical forest, which stretched far away on every side; the rude, uncultured savages who gathered round me—all had their influence in determining the emotions with which I gazed upon this 'thing of beauty.' I thought of the long ages of the past, during which the successive generations of this little creature had run their course—year by year being born, and living and dying amid these dark and gloomy woods, with no intelligent eye to gaze upon their loveliness; to all appearance such a wanton waste of beauty. Such ideas excite a feeling of melancholy. It seems sad that on the one hand such exquisite creatures should live out their lives and exhibit their charms only in these wild, inhospitable regions, doomed for ages yet to come to hopeless barbarism; while on the other hand, should civilized man ever reach these distant lands, and bring moral, intellectual, and physical light into the recesses of these virgin forests, we may be sure that he will so disturb the nicely-balanced relations of organic and inorganic nature as to cause the disappearance, and finally the extinction, of these very beings whose wonderful structure and beauty he alone is fitted to appreciate and enjoy. This consideration must surely tell us that all living things were *not* made for man."

To transpose what Wallace articulated in nineteenth-century prose so that his words can be applied usefully and inspiringly at the turn of the twenty-first century does not take much imagination, just willingness—on the part of Indonesians and the rest of humankind, together.

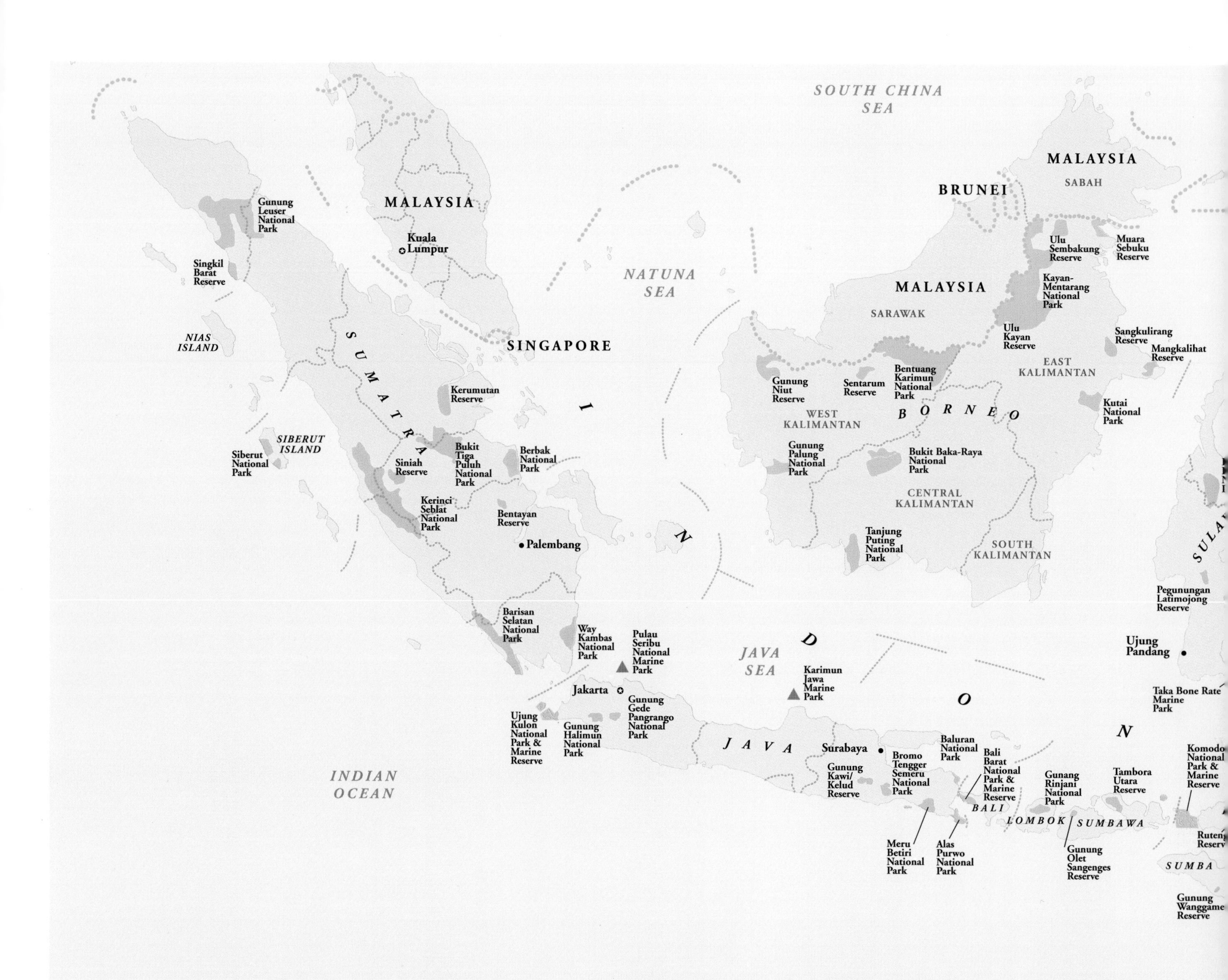

The national parks, major nature reserves, and marine reserves of Indonesia. Small dotted lines show Indonesian provincial boundaries; large

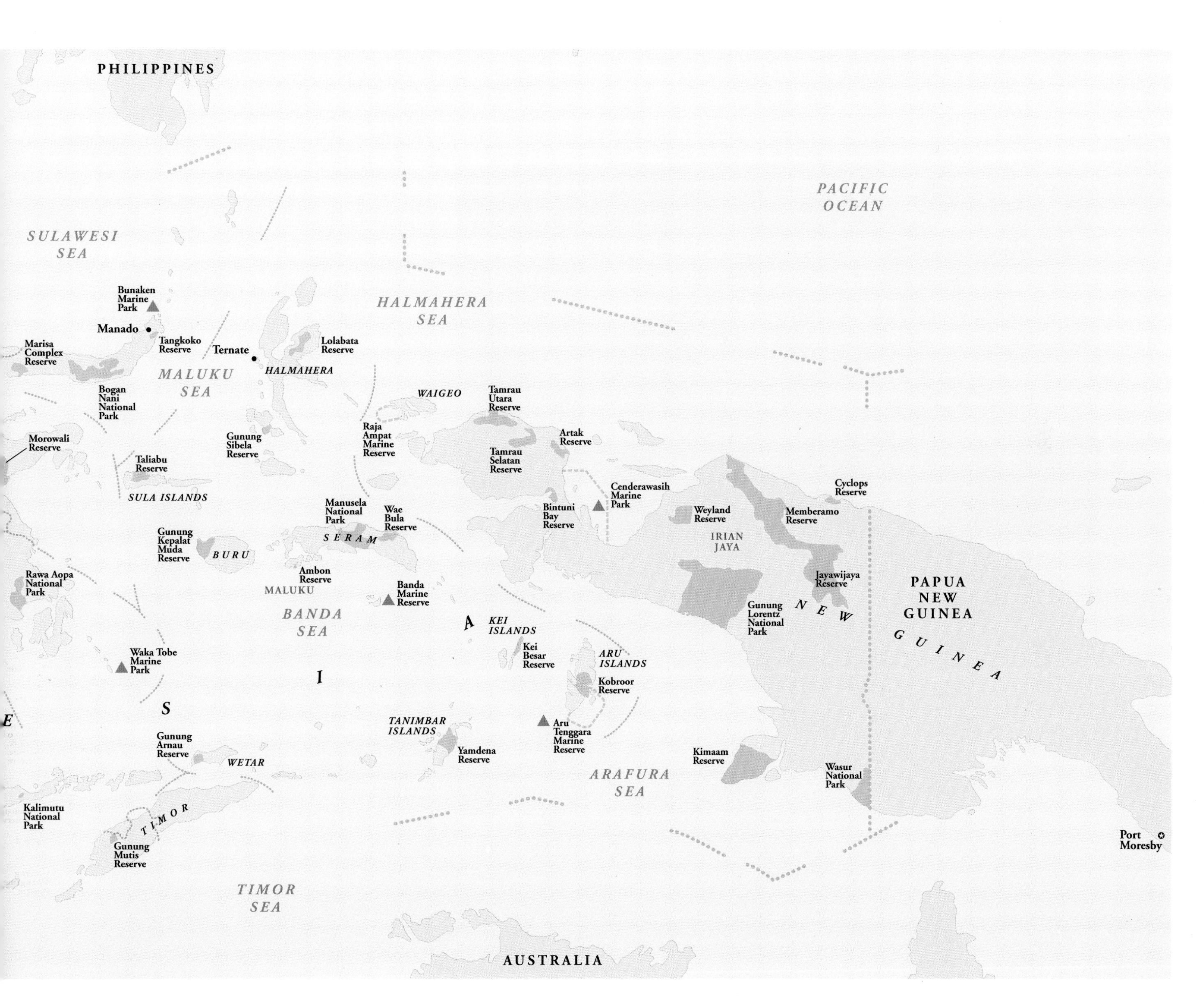

dotted lines show national boundaries.
Note: modern political English place-names are used.

Conservation of Indonesia's Biodiversity

A bustling bird market in Jakarta. Many species are captured illegally from their forest habitat for the pet trade, but the overwhelming majority of the birds die before reaching their destination. (James Martin)

Humans have long left their imprint on the Indonesian archipelago. Pleistocene hunter-gatherers subsisted on the bounties of rain forests and reefs. About twelve thousand years ago, during the Mesolithic period, humans domesticated plants and animals and reaped the first harvests. Three thousand years ago, people from Asia introduced the accoutrements of their culture: pottery, bronze, irrigation, and urbanization. More than a thousand years ago Chinese and Arab traders recognized the abundance of the archipelago's natural treasures of spices, sandalwood, ivory, and animals, and advanced technologies and religions were exchanged. By the sixteenth century, Westerners had discovered the archipelago's cornucopia, and colonization began in earnest with the establishment of plantations for rubber, tea, coffee, and spices. Yet only in the past hundred years has human activity begun to take its toll on the wildlands and seas of Indonesia. Today, pressures exerted by a swelling population already in excess of 206 million threaten to extinguish much of the archipelago's rich biological diversity—the evolutionary productions of hundreds of millions of years—in the blink of an eye, geologically speaking.

"Although Indonesia covers only 1.3 percent of the earth's land surface, it harbors 15 to 25 percent of the world's species," says Effendy Sumardja of Indonesia's Ministry of Environment. "Over 6,000 plant and animal species are used on a daily basis; 7,000 fish species provide a major protein source for the country, and more than a hundred species of rain forest trees are harvested commercially. Active stewardship of these resources is critical—

it is in Indonesia's long-term economic, social, and political interest."

Indonesia's remaining forested area, which covers 143 million hectares, is being cleared at an alarming rate: over 1 million hectares a year. Commercial logging is now the country's second-largest income earner. Like gashes in the earth's flesh, logging roads have opened even the remotest areas of Kalimantan and Irian Jaya to immigrant settlements that further erode the forest. Whereas the shifting agricultural practices of traditional forest people, who maintained low population densities, left little or no impact on forests, these practices have given way to transmigration schemes and large-scale conversion to monocultures of food and cash crops like rice, coffee, rubber, and oil palm. Already most of Indonesia's lowland rain forest, one of the most species-rich ecosystems on earth, is gone.

The devastation extends to the archipelago's coastal areas, where most of the population is concentrated. Expansive tracts of mangroves, once the world's largest, have been converted to shrimp farms. Riverine systems have succumbed to pollution from industry and agriculture. Coastal reefs and fisheries have been overcome by sedimentation as soils erode from deforested lands upstream, or by coral mining and destructive fishing methods.

The impact of disturbance to Indonesia's wildlands reverberates around the world. Indonesia now has the world's longest list of species threatened with imminent extinction, including 126 birds, 63 mammals, and 21 reptiles. Wildlife contraband—orangutans from Kalimantan, black palm cockatoos from Irian Jaya, Sumatran rhino horn—appears in places far and wide, as animals

Unbroken forest still cloaks the rugged mountains of Lore Lindu National Park, an important watershed for surrounding agricultural valleys. Most of the endemic mammals of Sulawesi live within Lore Lindu, although many are shy and not easy to see. (Jez O'Hare)

Above left: Sulawesi tarictic hornbill, one of two endemic hornbill species in Sulawesi. Lore Lindu National Park provides a refuge for many of Sulawesi's endemic birds. (Tui de Roy / Roving Tortoise)

Above right: In many areas along the borders of Lore Lindu National Park, no buffer zone exists and the park's forests are in direct contact with a sea of encroaching agriculture. (Marty Fujita)

Left: Ancient stone statue of the Bada Valley in Central Sulawesi, a remnant of a mysterious megalithic tradition. A number of these statues and burial vats have been found in the area, with age estimates varying from 3000 B.C. to A.D. 1300. (Kal Muller / PhotoBank)

are wrenched from their home by unscrupulous traders and buyers in international networks.

The need to protect these precious resources was recognized at Indonesia's inception as a sovereign nation in its 1945 constitution. Conservation laws have been enacted, and an impressive system of protected areas is in place. International conventions and treaties have been ratified to stem illegal trade in wildlife, protect wetlands, and enact conservation strategies and tropical forestry plans. But are the forces of change—funding, commitment, political and public will—strong enough to buttress the natural resource foundation on which the country's successes have been built? What is the responsibility of the rest of the world in helping Indonesia safeguard its natural treasures?

Conservation programs in Indonesia have, as in many parts of the world, focused on establish-

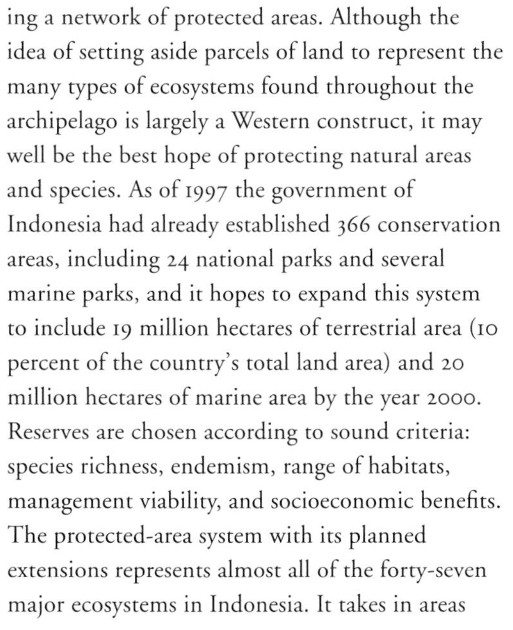

ing a network of protected areas. Although the idea of setting aside parcels of land to represent the many types of ecosystems found throughout the archipelago is largely a Western construct, it may well be the best hope of protecting natural areas and species. As of 1997 the government of Indonesia had already established 366 conservation areas, including 24 national parks and several marine parks, and it hopes to expand this system to include 19 million hectares of terrestrial area (10 percent of the country's total land area) and 20 million hectares of marine area by the year 2000. Reserves are chosen according to sound criteria: species richness, endemism, range of habitats, management viability, and socioeconomic benefits. The protected-area system with its planned extensions represents almost all of the forty-seven major ecosystems in Indonesia. It takes in areas where the largest number of native fauna can be found, and it exceeds the total area devoted to conservation in most other countries. Even so, problems and competing interests plague the system, such that today most of its reserves are little more than lines on maps.

All lands and waters are held by the state; traditional land rights, while recognized in theory, are ignored in practice, and creation of parklands has been, on paper at least, a simple matter of the national government establishing a boundary on the ground. The imposition of such boundaries, though, has often displaced indigenous people who have lived simply off the land for many generations. Conflicts arise with the displaced, who cannot understand why seemingly abstract conservation reasons for establishing the park (such as protection of watersheds, conserving a

Above left: Butterflies in a breeding cage at Lore Lindu National Park in Central Sulawesi. (Jez O'Hare)

Above right: Pupae of the butterfly *Papilio blumei.* A butterfly farming cooperative has been established in villages adjacent to Lore Lindu National Park to sustainably harvest several species for export to butterfly aviaries around the world. (Jez O'Hare)

Rattan fronds in a lowland forest in Sulawesi. Rattans are climbing palms with sharp, recurved spines that allow them to scale large trees to reach the sunlight above the canopy. (Jez O'Hare)

Above right: Rattan collector in Central Sulawesi. A significant portion of the rattan exported from Indonesia now comes from the forests of this province. (Jez O'Hare)

storehouse of genetic material for potential medicines, climate regulation) should be more important than their own survival. In many places, parks and people have become bitter enemies, and where once building materials, foodstuffs, and wildlife were harvested freely, these practices are now subject to penalties.

The greatest threats to established parks are huge infrastructure projects—dams, roads, logging concessions, and transmigration schemes—and in some cases, tourism. All bring in more people, more pressures, that erode the parks' integrity. Since parks are rarely, if ever, considered integral to local land-use planning, in part because of the ignorance of local and national authorities, development projects proceed with little regard for the parks' well-being. Woefully inadequate budgets, often amounting to no more than a few cents per hectare, and a lack of trained people to monitor and enforce conservation laws, patrol borders, and implement management plans render the park authority helpless. Boundaries are rarely safeguarded in even the most prestigious of parks.

Recognizing the political and economic constraints facing the traditional park concept, Indonesia is now trying new ideas on a grand scale, ideas born of the need to resolve conflicts between parks and people, conservation and development. Collectively called integrated conservation and development projects (ICDPs), this new approach aims to provide incentives that enhance the local benefits, and hence attractiveness, of conservation

The inimitable Komodo dragon, a carnivore that can reach lengths of up to 3 meters, is the world's largest lizard. It has a very restricted range but today finds refuge in Komodo National Park. (Roland Seitre / Peter Arnold, Inc.)

Left: Coastal ecosystems support many different forms of life, human included. The protection of marine resources, which are easily disturbed by human interference, is becoming increasingly important. (Michael Nichols)

Mount Leuser National Park in North Sumatra protects some of the best remaining tracts of lowland rain forest in the province, providing habitat for many of Sumatra's endangered species. (Jez O'Hare)

and sustainable resource use in and around parks. In essence, ICDPs strive to convey sound economic reasons for protecting a park from external forces. Individual projects can be exceedingly complex and ambitious, and Indonesia deserves credit for pioneering such a bold experiment, which is viewed by many as the only hope for the most important parks in Indonesia.

The ICDP approach now dominates conservation efforts in Indonesia, and funding from government and international agencies for these projects has more than doubled in recent years. Although the total amount ($200 million as of 1997) is small compared to the funds that pour into development or are reaped from natural resource exploitation, the very fact that money is being allocated for conservation signifies a growing awareness within Indonesia and among the world's developed countries of the urgency of preserving the archipelago's last great natural places.

More complete integration of parks into the larger landscape—one that includes people—is a key ingredient of the ICDP effort. Suraya Afiff, an environmental activist, says that "we have to examine each park individually—its size, shape, the number and degree of threats—and design our efforts to fit each situation. Most importantly, we need to examine the impact of conservation efforts on local people, and we need to empower them in these efforts. Conservation will not exist if it is not supported by local people, and we should find ways to work with them more effectively."

Some creative projects have begun to demonstrate these concepts. Lore Lindu National Park, established in 1982, is one of the largest remaining forested areas in Sulawesi. Its 231,000 hectares are home to almost all the endemic species of the island as well as ancient megaliths of historical and cultural significance. Villages surround the park, isolating it like an island in a sea of agriculture. Patches of the park are being cut down to grow coffee and cocoa for export; illegal harvesting of wildlife and rattan has diminished numbers of many species; and a planned road system and hydroelectric dam threaten the park's integrity. Nongovernmental organizations are working with park authorities, provincial government, and development agencies to integrate Lore Lindu into provincial land-use plans. Together with villagers who once felt alienated by the park, they are designing modest businesses: butterfly farming; a white-water rafting business employing former rattan poachers who routinely braved the rapids of the Lariang River to ferry rattan contraband to distant markets; sustainable rattan harvesting; and wild honey collection and marketing. The idea is to create income-generating activities that utilize the park's resources while promoting protection of habitat.

Komodo National Park is known around the world for its dragons, but less so for its incredible diversity of marine life, encompassed in 132,000 hectares of marine reserves. Several fishing villages lie within the park's boundaries, and many fishermen harvest marine resources in a destructive way. The park authority, working with nongovernmental organizations and villagers, has launched an awareness campaign aimed at the children of fisherfamilies. The idea is to help youngsters understand the negative consequences of dynamite fishing and the positive value of the park in maintaining the marine resources on which their families depend. Income-generating alternatives to destructive fishing practices are being introduced as well: ecotourism diving, mariculture projects for sea cucumbers and reef fish, and ways of marketing marine resources that cut out the middlemen who supply reef-destroying dynamite and cyanide to fishermen.

In Sumatra at Gunung Leuser National Park, a novel approach to park management has begun. Gunung Leuser contains some of the last vestiges of lowland rain forest in Sumatra and a huge diversity of Asiatic species, including orangutans, clouded leopards, gibbons, sun bears, and over 60 percent of all the bird species found on the island. As elsewhere, organized illegal logging, agricultural encroachment, wildlife poaching, and road construction all threaten the park. Recently, however, a private foundation with high-level government support and international funding has been granted the right to manage the 800,000-hectare park as well as almost a million hectares of the surrounding area. A partnership among the foundation, provincial government authorities, law enforcement agencies, universities, villages, and park managers has been established, and community development objectives—schools, roads, sanitation, potable water, health clinics—for the region will be met in exchange for better park protection. This is an undertaking of major proportions, a grand experiment that, if successful, could provide Indonesia with a radical new model for balancing conservation and development.

"The protected area system is important for protecting Indonesia's biodiversity," says Afiff. "But it is only one component of conservation. We have to look at strategies for entire provinces, and we have to build a conservation ethic into all levels of our society. It's not enough to have a few good people within the government committed to doing the right thing." Afiff and others of her generation are part of a rising tide of awareness within the country, working to ensure the future of Indonesia's unique natural heritage.

Environmental awareness programs strive to build a conservation ethic in Indonesia that will help preserve the country's natural heritage. (Jez O'Hare)

EPILOGUE

John C. Sawhill President and CEO, The Nature Conservancy

Gracefully arching between two continents, the Indonesian archipelago showcases some of the natural world's most spectacular creations: from 35-centimeter-high miniature deer to maleo fowl that incubate their eggs with geothermal heat, and from one-horned rhinos to the planet's largest blooming flower, the meter-wide rafflesia. Nature has been lavishly productive here. We now threaten to undo what the earth has taken eons to bring into existence.

Indeed, Alfred Russel Wallace would hardly recognize today's Indonesia. His beloved tropical forests are now full of the deafening noise of logging saws, and the rivers he once recorded are fouled by mine tailings. As the integration of global markets continues at a dizzying rate, the pace of the country's natural resource exploitation—often driven by the consumer demands of industrialized nations—has quickened. To many observers, scenes of modern-day Indonesia, with its dying coral reefs and ravaged rain forests, echo scenes from around the world. For far too many people, such environmental ruin has become accepted as an inevitable aspect of modernization. But as the fragility of the Southeast Asian economic "miracle" in recent years has shown, biodiversity conservation should not be subjected to the whims of the global economy; the biological treasures of this land deserve to be more than a mere footnote in economic history. As we close the door on the twentieth century, Indonesia's unique natural heritage must get the protection it merits.

On the eve of a new millennium, the globalization of communication, technology, and trade stands as a resounding testament to the power of human ambition and potential. Human ingenuity resonates around the globe as our inventions, ideas, and products, ranging from the silicon chip to the jet engine, shrink political borders and connect us to what were formerly the most remote corners of the planet. Perhaps less obvious—yet no less important—is the global impact these same activities are having on our common natural heritage. The growing market for tropical hardwoods, for instance, has been driven by an insatiable demand in the United States, Europe, and Japan. As a result, commercial logging now represents the greatest threat to Indonesia's lowland rain forests, the world's richest habitat for biodiversity. With such powerful new economic forces and technological advances in place, people can now alter the environment on a truly international scale. Reconciling this brave new world with the natural world that ultimately sustains it promises to be our greatest challenge in the next century.

Nowhere is this challenge more evident than in Indonesia. Since the time of Alfred Russel Wallace, Indonesians have benefited from the fruits of industrialization and inclusion in world markets. But the country also suffers from some of the worst kinds of environmental abuse and neglect, and this environmental degradation has the capacity to impoverish not only the Indonesian people but the rest of the world as well.

Like many plant and animal species across the globe, Indonesia's immense diversity of life faces increasingly complex and destructive threats. The decimation of biodiversity in Indonesia, and throughout the tropics, may bring heretofore unseen ecological devastation. If entire systems unravel, we will threaten the very roots of human prosperity. Already the unprecedented fires that raged throughout Indonesia because of unsustainable logging and clearing practices have destroyed critical wildlife habitat and endangered the health of tens of millions of people.

Of course, the economic and utilitarian value of biodiversity is hardly limited to the maintenance of basic ecosystem services. For instance, we have barely begun to realize the medicinal potential of nature. A full 25 to 40 percent of the drugs prescribed in the United States contain active ingredients derived from plants—and all this after we have analyzed less than 5 percent of the globe's plant species for their pharmaceutical possibilities. Indigenous peoples have long used plants for medicinal purposes, and modern science has finally started to acknowledge how much we can learn from local medical applications.

Nature's genetic resources also remain integral to commercial agriculture production. As much as 80 percent of the world food supply depends on fewer than two dozen species of plants and animals—a lack of genetic variation that leaves us vulnerable to drought and disease. Boosting crop production and ensuring crop productivity, so necessary for feeding growing human populations like that of Indonesia, require the diversity of species that we now squander in our rush to modernize.

One attempt to halt such wholesale destruction of species diversity involves local sustainable development projects. In Indonesia, for example, efforts to work with local communities to harvest and sell rattan, a climbing palm highly valued in furniture markets, hold hope for protecting rattan species throughout Sulawesi's forests. By providing these types of economic incentives to conserve natural resources, we may manage to stem the tide of habitat loss and extinction.

The need for such programs has never been more acute. Today's spasm of species extinction, unparalleled since the mass extinctions of dinosaurs millions of years ago, has unknown consequences, not only for Indonesia but for all of us. Although we might try to imagine what we lose with each species that disappears, we will never know for certain. As species unfamiliar to science silently perish, we lose yet more opportunities to discover a blight-resistant crop or a cure for an intractable disease. Pure self-interest alone should force us to pay attention to the ongoing loss of biodiversity, for species extinction is irreversible. Once gone, a species can never return.

Long before there was ever a global economy, there was a global ecology—a web of life imbuing species and ecosystems with a common future. The laws of this natural world have remained constant throughout time; human activity cannot escape its rules and logic. No longer can we ignore our role in creating the current epidemic of species extinction and its ravaging effects on our global environment. From Indonesia to our own backyards, wherever they may be, leaving an intact, thriving natural heritage for future generations remains our most urgent responsibility.

APPENDIX OF COMMON AND SCIENTIFIC NAMES

BIRDS

barbet	*Megalaima* spp.
barbet, gold-whiskered	*Megalaima chrysopogon*
bee-eater, purple-bearded	*Meropogon forsteni*
bee-eater, rainbow	*Merops ornatus*
bird of paradise, greater	*Paradisaea apoda*
bird of paradise, king	*Cicinnurus regius* (*Paradisea regia* in Wallace)
bird of paradise, King of Saxony	*Pteridophora alberti*
bird of paradise, Lawe's Parotia	*Parotia lawesii*
bird of paradise, lesser	*Paradisaea papuana*
bird of paradise, Raggiana (Marquis de Raggio bird of paradise)	*Paradisaea raggiana*
bird of paradise, red	*Paradisaea rubra*
bird of paradise, superb	*Lophorina superba*
bird of paradise, Wallace's standard-wing	*Semioptera wallacii*
broadbill, black and yellow	*Eurylaimus ochromalus*
broadbill, green	*Calyptomena viridis*
bulbul	family Picnonotidae
cassowary, Bennett's (dwarf cassowary)	*Casuarias bennetti*
cockatoo, black palm	*Probosciger aterrimus*
cuckoo shrike, cerulean	*Coracina temminckii*
flowerpecker	*Dicaem* spp.
goshawk, spot-tailed	*Accipiter trinotatus*
hawk eagle, Javan	*Spizaetus bartelsi*
honeyeater, crimson	*Myzomela dipapha*
hornbill	family Bucerotidae
hornbill, red-knobbed	*Rhyticeros cassidex*
hornbill, rhinoceros	*Buceros rhinoceros*
hornbill, Sulawesi tarictic	*Penelopides exarhatus*
kingfisher, blue-eared	*Alcedo meminting*
kingfisher, racquet-tailed	*Tanysiptera galatea nais*
lorikeet, ornate	*Trichoglossus ornatus*
lory, red (crimson lory)	*Eos bornea*
maleo (Celebes brush turkey)	*Macrocephalon maleo*
malkoha, yellow-billed	*Phaenicophaeus calyorhynchus*
mynah, fiery-browed	*Enodes erythrophris*
parrot, blue-backed	*Tanygnathus sumatranus*
parrot, eclectus	*Eclectus roratus*
parrot, hanging	*Loriculus* spp.
pheasant, Argus	*Argusiana argus*
pheasant, fireback	*Lophura ignita*
pigeon, crown	*Goura* spp.
pigeon, fruit	*Ducula* spp.
pigeon, Nicobar	*Caloenus nicobarica*
racquet tail, golden-mantled	*Prioniturus platurus*
roller, purple-winged	*Coracias temminckii*
starling, Bali	*Leucopsar rothschildi*
sunbird, black	*Nectarinia aspasia*
sunbird, crimson	*Aethopyga siparaja*
toucan	family Ramphastidae
woodswallow, ivory-backed	*Artamus monachus*

MAMMALS

anoa	*Bubalus depressicornis*
babirusa	*Babyrousa babyrussa*
bat, dawn	*Eonycteris spelaea*
bat, fruit (flying fox)	*Pteropus vampyrus*
bat, Horsfield's fruit	*Cynopterus horsfieldi*
bat, nectar	*Eonycteris spelaea*
bat, short-nosed fruit	*Cynopterus brachyotis*
bat, Wallace's stripe-faced	*Styloctenium wallacei*
binturong	*Arctictis binturong*
chimpanzee	*Pan troglodytes*
civet, banded palm	*Hemigalus derbyanus*
civet, Malay	*Viverra tangalunga*
civet, Sulawesi giant	*Macrogalidia musschenbroekii*

colugo (flying lemur)	*Cynocephalus volans*
cuscus, bear	*Ailurops ursinus*
cuscus, spotted	*Phalanger maculatus*
cuscus, Sulawesi dwarf	*Strigocuscus celebensis*
deer, miniature	*Tragulus javanicus*
elephant, Asian	*Elephas maximus*
flying squirrel	subfamily Petauristidae
gibbon	*Hylobates* spp.
gibbon, Bornean	*Hylobates muelleri*
gorilla	*Gorilla gorilla*
honey glider	*Petaurus breviceps*
leaf monkey or langur	*Presbytis* spp.
leaf monkey, banded	*Presbytis melalophos*
leaf monkey, red	*Presbytis rubicunda*
leopard, clouded	*Neofelis nebulosa*
macaque, crested black	*Macaca nigra*
macaque, long-tailed	*Macaca fascicularis*
macaque, pig-tailed	*Macaca nemestrina*
monkey, proboscis	*Nasalis larvatus*
orangutan	*Pongo pygmaeus*
pangolin	*Manis javanica*
phalanger	*Phalangeri* spp.
pig, bearded	*Sus barbatus*
rhinoceros, Sumatran	*Dicerorhinus sumatrensis*
siamang	*Hylobates syndactylus*
squirrel, giant	*Ratufa affinis*
squirrel, pygmy	*Exilisciurus exilis*
sunbear	*Helarctos malayanus*
tarsier	*Tarsius bancanus*
tarsier, spectral	*Tarsius spectrum*
tiger, Sumatran	*Panthera tigris*
tree kangaroo, Goodfellow's	*Dendrolagus goodfellowi*

REPTILES AND AMPHIBIANS

frog, flying	*Rhacophorus palmatus*
frog, horned	*Megophrys nasuta*
frog, tree	*Rhacophorus pardalis*
Komodo dragon	*Varanus komodoensis*
lizard, Bornean crested	*Calotes cristatellus*
lizard, flying	*Draco* spp.
python, green tree	*Morelia viridis* (syn. *Chondropython viridis*)
turtle, green sea	*Chelonia mydas*
turtle, leatherback	*Dermochelys coriacea*

FISH

anemone fish, pink	*Amphiprion perideraion*
angelfish	family Chaetodontidae
damselfish, magenta	family Pomacentridae
grouper, coral	*Cephalopholis miniata*
lionfish	*Pterois volitans*
shark, hammerhead	*Sphyrna* spp.
wrasse, Napoleon (humphead wrasse)	*Cheilinus undulatus*

MARINE INVERTEBRATES

anemone shrimp	*Periclimenes* spp.
coconut crab	*Birgus latro*
feather star crinoid	family Comatulidae
goby shrimp	family Alpheidae
jellyfish	*Mastigius* spp.

INSECTS

beetle, tortoise	family Chrysomelidae
butterfly (Brazilian)	*Callithea leprevri*
butterfly (Sulawesi)	*Papilio blumei*
butterfly, archduke	*Lexius pardalus*
butterfly, birdwing	*Ornithoptera* spp., including *O. brookeana, O. croesus, O. goliath, O. poseidon; Trogonoptera* spp.; *Troides* spp.
butterfly, Sumatran (Indian leaf)	*Kallima paralekta amplirufa*
butterfly, swallowtail	*Graphium androcles*
butterfly, swallowtail (blue triangle)	*Graphium sarpedon*
butterfly, Ulysses (blue emperor)	*Papilio ulysses*

fig wasp	family Agaonidae; *Waterstoniella masii* in photo
grasshopper, short-horned	family Acrididae
katydid	family Tettigoniidae
leaf insect	family Phyllidae
walking stick	family Phasmidae

PLANTS

annonaceous trees	family Annonaceae
banana, wild	*Musa* spp.
clove	*Syzygium aromaticum*
dipterocarp	*Shorea* spp., *Dipterocarpus* spp.
durian	*Durio zibenthinus*
ebony	*Diospyros celebica*
eucalyptus	*Eucalyptus deglupta*
koompassia	*Koompassia malaccensis*
mango	*Mangifera indica*
mangrove	*Rhizophora* spp.
mushroom, phosphorescent	*Mycena* spp.
nutmeg	*Myristica fragrans*
orchid, Irian Jaya	*Dendrobium smilliae*
orchid, Kalimantan black	*Coelogyne pandurata*
palm, fishtail	*Caryota mitis*
palm, lontar	*Borassus flabellifer*
palm, Sulawesi endemic	*Pigafetta filaris*
palm, woka	*Livistonia rotundifolia*
pitcher plant	*Nepenthes stenophylla*
rafflesia	*Rafflesia arnoldi*
rattan	*Calamus* spp.
strangler fig	*Ficus caulocarpa*

NOTES

1. THE EVOLUTION OF A NATURALIST

Wallace's own account of his life, written in his old age, is in Wallace 1905; for published correspondence and other autobiographical material, see Wallace 1916. For a range of biographical perspectives, see George 1964; Williams-Ellis 1966; McKinney 1972; Brackman 1980; Fichman 1981; Clements 1983; Brooks 1984. A useful short essay is John Bastin's introduction to his edition of *The Malay Archipelago* (Kuala Lumpur: Oxford University Press, 1986).

Page

4 "Central and controlling incident": Wallace 1905, 1:336.

5f. Self-appraisal: Wallace 1905, 1:33, 179, 224, 225; *Independent,* March 9, 1899, 668.

6 Phrenology: Wallace 1905, 1:25, 26, 224, 235, 159; George 1964, 239; Jardine, Secord, and Spary 1996, 348.

6f. Wallace's nature rambling and botanizing: Wallace 1905, 1:109, 110, 190, 193.

7f. British enthusiasm for nature: Allen 1976; Barber 1980; Browne 1983; Lloyd 1985; Merrill 1989.

9 Shows and exhibits: Altick 1978. Fuegians: Desmond and Moore 1991, 147; Browne 1995, 234–235. World traffic in nature: Grove 1995; Jardine, Secord, and Spary 1996, 147, 343–349; Browne 1983, 77ff.; MacKenzie, 1990.

10f. New information in conflict with old understandings: Browne 1983, 5, 16; Barber 1980, 65–68, 76; Quammen 1996, 34.

13 Darwin's reluctance to publish: Desmond and Moore 1991.

13f. Wallace's reading: George 1964, 2; McKinney 1972, 6; Quammen 1996, 58–59; Camerini 1987; Hodge 1991.

14 "Permanent possession": Wallace 1905, 1:232. Wallace meets Henry Bates: Wallace 1905, 1:237ff. "Great number and variety . . .": Wallace 1905, 1:237. "Rather a wild scheme": Wallace 1905, 1:254. The Amazon: Edwards 1847. "Making collections in Natural History": Wallace 1905, 1:264ff.

16 Charles Waterton: Barber 1980, 99. Preparing for the Amazon expedition: Wallace 1905, 1:265–267; Brackman 1980, 128.

16ff. In the Amazon: Wallace 1895; Wallace 1905, 1:267ff.
17 "Perfect nudity of these daughters of the forest": Wallace 1853a, 204.
20 The importance of geography for Wallace: George 1964, 24; George 1979, 505; Camerini 1987.
20f. Return home and shipwreck: Wallace 1895, 271ff.
21 "Fifty times since I left Pará . . .": quoted in Quammen 1996, 71. In London after the Amazon: Wallace 1905, 1:313, 324.

2. PREPARING FOR THE ARCHIPELAGO

Page
27 "If we look at a globe . . . ," "to the ordinary Englishman . . .": Wallace [1869a] 1962, chap. 1, 1. "One of the chief volcanic belts upon the globe": Wallace [1869a] 1962, chap. 1, 4. "Bathed by the tepid water . . .": Wallace [1869a] 1962, chap. 1, 1.
28 Southeast Asian specimens in Britain: Wallace 1905, 1:327–328; Scherren 1905, 65. The babirusa: Belcher 1848, 2:118.
31 Orangutans: Schwartz 1987, 1988; Harrisson 1987; Nadler et al. 1995. Live orangutans and specimens in England: *Murray's Encyclopedia* 1837, 2:506; Desmond 1989, 291; Desmond and Moore 1991, 243–244, 263, 287, 593. Wallace's interest in orangutans: Wallace 1916, 1:53.
31ff. Birds of paradise: Gilliard 1969; Beehler 1989, 1991; Lesson 1835.
33f. Voyage to the archipelago: Wallace 1905, 1:330ff.
34ff. Wallace's travels among the islands: see John Bastin's introduction to *The Malay Archipelago* (Kuala Lumpur: Oxford University Press, 1989).
34 Terrors of the London streets: George 1964, 48.
34f. Wallace's contacts in the field: Camerini 1996.
35 Equipment: George 1979.
36 *Ung-lung:* Wallace [1869a] 1962, chap. 31, 351. Wallace collecting as a hilarious spectacle: Wallace 1858a, 6122. "One day when I was rambling . . .": Wallace [1869a] 1962, chap. 29, 325. "A few years before . . .": Wallace [1869a] 1962, chap. 31, 348–349. "Wherever I went . . .": Wallace [1869a] 1962, chap. 15, 171.

Wallace's Biological Laboratory

Quotations are from written responses to interview questions from James Moore (Dec. 1996) and Dan Simberloff (Sept. 1996).

3. ENCOUNTERING THE ORANGUTAN

Page
42 Wallace on Borneo: Wallace [1869a] 1962, chaps. 4–5.
43 "Its consistence and flavour . . . ," "a most disgusting odour," "Trees and fruits . . .": Wallace [1869a] 1962, chap. 5, 57, 58.

47ff. Wallace and orangutans: Wallace 1856a; Wallace 1856b; Wallace [1869a] 1962, chap. 4.
47 "Obtain good specimens . . .": Wallace [1869a] 1962, chap. 4, 30. Shooting orangutans in Borneo: Harrisson 1987, 17, 19–21.
47f. The baby orangutan: Wallace 1856c; Wallace [1869a] 1962, chap. 4, 32ff.
48 "It was a never-failing source of amusement . . .": Wallace [1869a] 1962, chap. 4, 34. "Skeleton dry & skin in arrack": Wallace, MS notebook, Linnean Society.
52 "Died" and "Killed . . .": Wallace [1869a] 1962, chap. 4, 35; Wallace, MS notebook, Linnean Society.
52f. Wallace's Sarawak paper: Wallace 1855.
52 "Surrounded by wild nature and uncultured man," "During the evenings . . .": Wallace 1905, 1:354–355.
53 Remarks heard by Samuel Stevens: Wallace 1905, 1:355.

Evolution's Panoply

Quotations are from taped and written responses to interview questions from Peter Ashton (Dec. 1996), Tim Laman (July 1997), and Kuswata Kartawinata (Jan. 1997).

The Wild Man of the Forest

Quotations are from written responses to interview questions from John Mitani (Sept. 1996) and Cheryl Knott (Jan. 1998).

4. BALI, LOMBOCK, AND CELEBES

Page
75 "The relations of animals . . ," "I have set myself . . .": Clements 1983, 40. "Commonly reckoned by weeks and months": Wallace [1869a] 1962, chap. 1, 1–2.
76 Wallace on Bali and Lombock: Wallace [1869a] 1962, chaps. 10–11. Megapodiidae: Wallace [1869a] 1962, chap. 10, 120.
79 Wallace's Line: the literature is enormous; see Mayr 1945; Fichman 1977; Whitmore 1981; Keast 1983; Camerini 1987, 1993.
79f. Celebes: Wallace 1860b; Wallace [1869a] 1962, chaps. 15–18.

Sulawesi: Island Enigma

Quotations are from written responses to interview questions from Duncan Neville (Jan. 1996), Chris Wemmer (Aug. 1996), and Steve Siebert (Nov. 1996) and from a telephone interview with Nora Bynum (Feb. 1998).

5. THE MOLUCCAS

Page

99 "The clearness of the water . . .": Wallace [1869a] 1962, chap. 20, 226.

104 Dutch ichthyologist: Wallace [1869a] 1962, chap. 20, 231.

106 Collecting on Amboyna: Wallace 1858a. "Much to be wished . . .": Wallace [1869a] 1962, chap. 20, 233. "Very rich . . . ," "ample space . . . ," "In this house . . .": Wallace [1869a] 1962, chap. 21, 234, 235–236.

107f. Wallace on Gilolo: over the years, Wallace put on paper a number of slightly varying accounts of how he came to his insight; see McKinney 1972, 80–81, 160–163. Scholarly and polemic discussion has been extensive; see Beddall 1968, 1972, 1988b; Brackman 1980; Brooks 1984; Kottler 1985; Hodge 1991; Quammen 1996; Moore 1997.

110 "Out with the theory . . .": quoted in Browne 1995, 541.

111 Wallace's paper presented to the Linnean Society: Wallace 1858b.

Marine Biodiversity

Quotations are from written responses to interview questions from Soekarno (Feb. 1997) and Rili Djohani (Feb. 1997) and from a telephone interview with Bob Johannes (Oct. 1997).

Darwin, Wallace, and Precedence

Quotations and impressions are from written and taped responses to interview questions from James Moore (Dec. 1996), Jonathan Hodge (Jan. 1997), David Quammen (Aug. 1997), Ernst Mayr (Oct. 1996), and Jane Camerini (Oct. 1996, with additional correspondence Nov. 1996–Sept. 1998).

6. THE ISLANDS OF KÉ, ARU, AND NEW GUINEA

Page

131 Wallace's voyage to the Aru Islands: Wallace [1869a] 1962, chap. 28; Wallace 1857b. "I was much delighted . . .": Wallace [1869a] 1962, chap. 29, 326.

131f. "But few European feet . . .": Wallace [1869a] 1962, chap. 28, 317.

135 Wallace at Aru: Wallace [1869a] 1962, chap. 30. "One of the most magnificent insects . . . ," "I trembled with excitement . . . ," and "quite another thing . . .": Wallace [1869a] 1962, chap. 30, 328–329.

136 "Which repaid me . . . ," "It was a small bird . . .": Wallace [1869a] 1962, chap. 31, 338–339. "The ornamental plumes . . .": Wallace [1869a] 1962, chap. 31, 349.

140 "Rejoice with me . . .": Wallace 1857–1858. "At early morn . . .": Wallace [1869a] 1962, chap. 31, 340–341.

143 "Their accounts . . .": Wallace 1862. Wallace at Dorey: Wallace [1869a] 1962, chap. 34; Lesson 1835.

144ff. Wallace on Batchian: Wallace [1869a] 1962, chap. 24; Wallace 1859; Wallace 1858–1859.

144 "I saw a bird . . .": Wallace [1869a] 1962, chap. 24, 252. "Here I have been . . .": Wallace 1859, 129. "A perfectly new and magnificent species . . .": Wallace [1869a] 1962, chap. 24, 257–258.
146ff. Wallace at Waigiou: Wallace [1869a] 1962, chap. 36.
147 "Two long rigid glossy ribands": Wallace [1869a] 1962, chap. 36, 402. "Here I lived . . .": Wallace [1869a] 1962, chap. 36, 406–407.
149 "Thus ended my search after these beautiful birds": Wallace [1869a] 1962, chap. 38, 439.
150ff. Bird of paradise family: Gilliard 1969; see also Elliot 1977, 1978; Cooper 1979; Beehler 1989.

Ecology and Behavior of Birds of Paradise

Quotations are from written responses to interview questions from Bruce Beehler (Nov. 1996).

7. JAVA, SUMATRA, AND HOME: REALIZATIONS

Page
161 "I cannot now put up so well . . .": Wallace 1861; George 1964, 46. Wallace on Java: Wallace [1869a] 1962, chap. 7.
161f. Wallace on Sumatra: Wallace [1869a] 1962, chap. 8.
162 "After crossing a stream . . .": Wallace [1869a] 1962, chap. 8, 106.
167 "I lost patience . . .": Wallace [1869a] 1962, chap. 8, 103.
167f. Bringing live birds of paradise to England: Wallace 1905, 1:383–384; Wallace [1869a] 1962, chap. 38, 423.
168 Birds of paradise in London: [anon.] 1862a; [anon.] 1862b; Scherren 1905. "I feel sure . . .": Wallace [1869a] 1962, chap. 38, 424.
168f. Wallace's collecting statistics: Bastin 1986, xvii.
169 "Regular species monger" . . . : Wallace 1905, 1:403. Wallace's publishing record: Smith 1991. "Not only as a token . . .": Wallace [1869a] 1962, v.
169ff. Wallace's life after returning to England; Wallace 1905, vol. 2.
171 "An accurate knowledge . . .": Wallace 1864.
173 "Now, though I always liked surveying . . .": Wallace 1905, 1:368–369. Wallace's finances: Wallace 1905, 2:377. Civil pension: Colp 1992.
174 Range of Wallace's thinking and writing: Smith 1991. Honors: George 1964, 279–286.
175 Thoughts on the United States: Wallace 1905, 2:190–196. Letter from Aru: George 1964, 40.
175f. Remembering Ali: Camerini 1996.
176 Wallace on life in the tropics: Wallace 1899.
176f. Linnean Society medal speech: Wallace 1908.
176 Wallace at Darwin's funeral: Wallace 1905, 2:102; Desmond and Moore 1991, 669.
177ff. Plume trade: Doughty 1975; Swadling 1996; *Popular Science Monthly,* Sept. 1874, 557; *Times,* Dec. 25, 1897; *Punch,* May 14, 1892.
177 "We shall be, indeed, astonished . . .": *Times,* May 16, 1862.
178 "It is possible . . .": *Murray's Magazine,* Jan.–June 1889, 375.
180 Rise of conservationist sentiment: Allen 1976, 198ff. Modern ecological study of birds of paradise: Beehler 1989, 1991.

181 Western man and natural environments: Browne 1983; MacKenzie 1990; Grove 1995; Jardine, Secord, and Spary 1996.

8. FROM THE MALAY ARCHIPELAGO TO INDONESIA: SPANNING THE CENTURIES

Page

185ff. Data on contemporary Indonesia: Whitten, Soeriaatmadja, and Afiff 1996; K. MacKinnon et al. 1996; Indonesia, Ministry of State for Population and Environment, 1992; Indonesia, Ministry of National Development Planning, 1993.

185 Impact of early human settlement and land use: Reid 1995.

196ff. Modern fieldwork: quotations are from interviews and written correspondence with working researchers. On the Sulawesi civet cat: Chris Wemmer; on thrip pollination of Bornean dipterocarps: Peter Ashton; on the Wana of Morowali: Michael Alvard; an Indonesian marine biologist's stories from the field: Rili Djohani; on rattan ecology: Steve Siebert; on the number of conservation biologists in Indonesia: Kuswata Kartawinata. General impressions of modern fieldwork come from these sources and from conversations with Tony Whitten, John Mitani, Nora Bynum, Cheryl Knott, Ernst Mayr, Marty Fujita, and Tim Laman.

201ff. The orangutan in difficulties: J. Mackinnon 1974; Maple 1980; de Boer 1982; Schwartz 1987; Kaplan and Rogers 1994; Nadler et al. 1995.

207ff. The 1997 fires and their aftermath: the day-by-day piling up of events generated hundreds upon hundreds of informational and analytical postings on the Internet, which are summarized here. See, for example, postings of articles by Peter Waldman in the *Wall Street Journal,* October 25 and 28, 1998.

217 "I knew how few Europeans . . .": Wallace [1869a] 1962, chap. 31, 339–340.

Conservation of Indonesia's Biodiversity

Quotations are from a taped response to interview questions from Effendy Sumardja (Apr. 1997), a telephone interview with Suraya Afiff (Feb. 1998), and correspondence with Michael Wells (Dec. 1996).

SELECTED BIBLIOGRAPHY

Allen, David Elliston. 1976. *The Naturalist in Britain: A Social History*. London: A. Lane.

Altick, Richard D. 1978. *The Shows of London.* Cambridge, Mass.: Harvard University Press, Belknap Press.

[Anon.]. 1862a. "Arrival of Living Birds of Paradise." *Times,* April 2, 9.

[Anon.]. 1862b. "The Birds of Paradise in the Zoological Society's Garden." *Illustrated London News,* April 12, 375.

Barber, Lynn. 1980. *The Heyday of Natural History, 1820–1870*. London: J. Cape.

Bastin, John. 1989. Introduction to Alfred Russel Wallace, *The Malay Archipelago: The Land of the Orang-utan and the Bird of Paradise. A Narrative of Travel, with Studies of Man and Nature.* Kuala Lumpur: Oxford University Press.

Beddall, Barbara G. 1968. "Wallace, Darwin, and the Theory of Natural Selection: A Study in the Development of Ideas and Attitudes." *Journal of the History of Biology* 1 (2): 261–332.

———, ed. 1969. *Wallace and Bates in the Tropics: An Introduction to the Theory of Natural Selection, Based on the Writings of Alfred Russel Wallace and Henry Walter Bates.* New York: Macmillan.

———. 1972. "Wallace, Darwin, and Edward Blyth: Further Notes on the Development of Evolutionary Theory." *Journal of the History of Biology* 5: 153–158.

———. 1988a. "Darwin and Divergence: The Wallace Connection." *Journal of the History of Biology* 21 (1): 1–68.

———. 1988b. "Wallace's Annotated Copy of Darwin's Origin of Species." *Journal of the History of Biology* 21 (2): 265–289.

Beehler, Bruce. 1989. "The Birds of Paradise." *Scientific American,* Dec. 1989: 17–123.

———. 1991. *A Naturalist in New Guinea*. Austin: University of Texas Press.

Belcher, Edward. 1848. *Narrative of the Voyage of H.M.S. "Samarang" during the Years 1843–1846.* London: Reeve, Benham & Reeve.

Brackman, Arnold C. 1980. *A Delicate Arrangement: The Strange Case of Charles Darwin and Alfred Russel Wallace.* New York: Times Books.

Brooks, John Langdon. 1984. *Just before the Origin: Alfred Russel Wallace's Theory of Evolution.* New York: Columbia University Press.

Browne, Janet. 1983. *The Secular Ark: Studies in the History of Biogeography.* New Haven: Yale University Press.

———. 1995. *Charles Darwin: A Biography.* Vol. 1: *Voyaging.* New York: Knopf.

———. 1996. "Biogeography and Empire." In *Cultures of Natural History,* edited by N. Jardine, J. A. Secord, and E. C. Spary, 305–321. New York: Cambridge University Press.

Camerini, Jane R. 1987. "Darwin, Wallace, and Maps." Ph.D. diss., University of Wisconsin–Madison.

———. 1993. "Evolution, Biogeography, and Maps: An Early History of Wallace's Line." *Isis* 84: 700–727.

———. 1996. "Wallace in the Field." *Osiris,* 2d ser., 11: 44–65.

Clements, Harry. 1983. *Alfred Russel Wallace: Biologist and Social Reformer.* London: Hutchinson.

Colp, Ralph, Jr. 1992. "'I will gladly do my best': How Charles Darwin Obtained a Civil List Pension for Alfred Russel Wallace." *Isis* 83 (1): 2–27.

Cooper, William T., with Joseph M. Forshaw. 1979. *The Birds of Paradise and Bower Birds.* Boston: David R. Godine.

Cubitt, Gerald S., Tony Whitten, and Jane Whitten. 1992. *Wild Indonesia: The Wildlife and Scenery of the Indonesian Archipelago.* Cambridge, Mass.: MIT Press, 1992.

Darwin, Charles. [1859] 1964. *The Origin of Species.* Cambridge, Mass.: Harvard University Press.

de Boer, Loebert E. M., ed. 1982. *The Orang utan: Its Biology and Conservation.* The Hague and Boston: W. Junk.

Desmond, Adrian J. 1989. *The Politics of Evolution: Morphology, Medicine, and Reform in Radical London.* Chicago: University of Chicago Press.

Desmond, Adrian, and James Moore. 1991. *Darwin.* New York: Warner Books.

Doughty, Robin W. 1975. *Feather Fashions and Bird Preservation: A Study in Nature Protection.* Berkeley: University of California Press.

Edwards, William H. 1847. *A Voyage up the River Amazon, Including a Residence at Pará.* New York: D. Appleton; Philadelphia: G. S. Appleton.

Elliot, Daniel Giraud. 1977. *A Monograph of the Paradiseida; or, Birds of Paradise.* New York: Johnson Reprint Corp.; Amsterdam: Theatrum Orbis Terrarum.

Everett, Michael. 1978. *The Birds of Paradise.* New York: Putnam.

Fichman, Martin. 1977. "Wallace, Zoogeography, and the Problem of Land Bridges." *Journal of the History of Biology* 10 (1): 45–63.

———. 1981. *Alfred Russel Wallace.* Boston, Mass.: Twayne.

George, Wilma. 1964. *Biologist Philosopher: A Study of the Life and Writings of Alfred Russel Wallace.* London: Abelard-Schuman.

———. 1979. "Alfred Russel Wallace, the Gentle Trader: Collecting in Amazonia and the Malay Archipelago, 1848–1862." *Journal of the Society for the Bibliography of Natural History* 9 (14): 503–514.

———. 1981. "Wallace and His Line." In *Wallace's Line and Plate Tectonics,* edited by T. C. Whitmore, 3–8. Oxford: Clarendon Press.

Gilliard, Thomas E. 1969. *Birds of Paradise and Bower Birds.* London: Weidenfeld & Nicolson.

Grove, Richard. 1995. *Green Imperialism: Colonial Expansion, Tropical Island Edens, and the Orgins of Environmentalism, 1600–1860.* Cambridge: Cambridge University Press.

Harrisson, Barbara. 1987. *Orang-utan*. New York: Oxford University Press.

Hodge, M. J. S. 1991. *Origins and Species: A study of the Historical Sources of Darwinism and the Contexts of Some Other Accounts of Organic Diversity, from Plato and Aristotle On*. New York: Garland.

Hudson, W. H. 1897. Letter to the editor about the trade in birds' feathers. *Times,* Dec. 25, 5.

Indonesia. Ministry of National Development Planning. 1993. *Biodiversity Action Plan for Indonesia.* Jakarta: Ministry of National Development Planning, National Development Planning Agency.

———. Ministry of State for Population and Environment. 1992. *Indonesian Country Study on Biological Diversity*. Jakarta: Ministry of State for Population and Environment.

Jardine, N., J. A. Secord, and E. C. Spary, eds. 1996. *Cultures of Natural History.* New York: Cambridge University Press.

Jepson, Paul. 1997. *Birding Indonesia: A Bird-Watcher's Guide to the World's Largest Archipelago.* Singapore: Periplus.

Kaplan, Gisela, and Lesley Rogers. 1994. *Orang-utans in Borneo.* Hanover, N.H.: University Press of New England.

Keast, J. Allen. 1983. "In the Steps of Alfred Russel Wallace: Biogeography of the Asian-Australian Zone." In *Evolution, Time, and Space: The Emergence of the Biosphere,* edited by R. W. Sims, J. H. Price, and P. E. S. Whalley, 368–407. New York: Academic Press.

Kottler, Malcolm J. 1974. "Alfred Russel Wallace, the Origins of Man, and Spiritualism." *Isis* 65: 145–92.

———. 1985. "Charles Darwin and Alfred Russel Wallace: Two Decades of Debate over Natural Selection." In *The Darwinian Heritage,* edited by David Kohn, 367–431. Princeton: Princeton University Press.

Lee, Tung-Yi, and Lawrence Lawver. 1995. "Cenozoic Plate Reconstruction of Southeast Asia." *Tectonophysics* 251: 85–138.

Lesson, René P. 1839. *Voyage autour du monde, entrepris par ordre du gouvernement, sur la corvette la Coquille.* Paris: P. Pourrat.

———. 1835. *Histoire naturelle des oiseaux de paradis et des epimaques.* Paris: A. Bertrand.

Lloyd, Clare. 1985. *The Travelling Naturalists.* London: Croom Helm.

MacKenzie, John M., ed. 1990. *Imperialism and the Natural World.* Manchester: Manchester University Press.

Mackinnon, John. 1974. "The Behaviour and Ecology of Wild Orang-utans" *Animal Behaviour* 23: 3–74.

MacKinnon, Kathy, G. Hatta, H. Halim, and A. Mangalik. 1996. *The Ecology of Kalimantan.* Hong Kong: Periplus.

Maple, Terry L. 1980. *Orang-utan Behavior.* New York: Van Nostrand Reinhold.

Mayr, Ernst. 1945. "Wallace's Line in the Light of Recent Zoogeographic Studies." In *Science and Scientists in the Netherlands Indies*, edited by Pieter Honig and Frans Verdoorn, 241–250. New York: Board for the Netherlands Indies, Surinam, and Curaçao.

———. 1982. *The Growth of Biological Thought: Diversity, Evolution, and Inheritance.* Cambridge, Mass.: Harvard University Press, Belknap Press.

McKinney, H. Lewis. 1972. *Wallace and Natural Selection.* New Haven: Yale University Press.

Merrill, Lynn L. 1989. *The Romance of Victorian Natural History.* New York: Oxford University Press.

Monk, Kathryn A., Yance de Fretes, and Gayatri Reksodiharjo-Lilley. 1997. *The Ecology of Nusa Tenggara and Maluku.* Hong Kong: Periplus.

Moore, James R., ed. 1989. *History, Humanity, and Evolution: Essays for John C. Greene.* Cambridge: Cambridge University Press.

———. 1994. *The Darwin Legend.* Grand Rapids, Mich.: Baker Books.

———. 1997. "Wallace's Malthusian Moment: The Common Context Revisited," in *Victorian Science in Context,* edited by Bernard Lightman, 290–311. Chicago: University of Chicago Press.

Nadler, Ronald D., Birute Galdikas, Lori K. Sheehan, and Norm Rosen, eds. 1995. *The Neglected Ape.* New York: Plenum Press.

Nelson, Gareth. 1978. "From Candolle to Croizat: Comments on the History of Biogeography." *Journal of the History of Biology* 11 (2): 269–305.

Quammen, David. 1996. *The Song of the Dodo: Island Biogeography in an Age of Extinctions.* New York: Scribner.

Reid, Anthony. 1995. "Humans and Forests in Pre-colonial Southeast Asia." In *Environment and History.* Cambridge: White Horse Press.

Scherren, Henry. 1905. *The Zoological Society of London: A Sketch of Its Foundation and Development and the Story of Its Farm, Museums, Garden, Menagerie and Library.* London: Cassell.

Schwartz, Jeffrey H. 1987. *The Red Ape: Orang-utans and Human Origins.* London: Elm Tree Books.

———, ed. 1988. *Orangutan Biology.* New York: Oxford University Press.

Sellato, Bernard. 1989. *Hornbill and Dragon.* Jakarta: Elf Aquitaine Indonésie.

Sheets-Pyenson, Susan. 1988. *Cathedrals of Science: The Development of Colonial Natural History Museums during the Late Nineteenth Century.* Kingston, Ont.: McGill–Queen's University Press.

Smith, Charles H. 1989. "Historical Biogeography: Geography as Evolution, Evolution as Geography," *New Zealand Journal of Zoology* 16: 773–785.

———, ed. 1991. *Alfred Russel Wallace: An Anthology of His Shorter Writings.* Oxford: Oxford University Press.

Stoddart, David R. 1994. "This Coral Episode: Darwin, Dana, and the Coral Reefs of the Pacific." In *Darwin's Laboratory: Evolutionary Theory and Natural History in the Pacific,* edited by Roy MacLeod and Philip F. Rehbock. Honolulu: University of Hawaii Press.

Swadling, Pamela. 1996. *Plumes from Paradise: Trade Cycles in Outer Southeast Asia and Their Impact on New Guinea and Nearby Islands until 1920.* Coorparoo, Queensland, Austr.: Papua New Guinea Nature Museum.

Tomascik, Tomas, A. J. Mah, A. Nontji, and M. K. Moosa. 1997. *The Ecology of the Indonesian Seas.* Vols. 1 and 2. Oxford: Oxford University Press.

Van Oosterzee, Penny. 1997. *Where Worlds Collide: The Wallace Line.* Ithaca, N.Y.: Cornell University Press.

Wallace, Alfred Russel. 1853a. *A Narrative of Travels on the Amazon and Rio Negro, with an Account of the Native Tribes, and Observations on the Climate, Geology, and Natural History of the Amazon Valley.* London: Reeve.

———. 1853b. *Palm Trees of the Amazon and Their Uses.* London: John Van Voorst.

———. 1855. "On the Law Which Has Regulated the Introduction of New Species." *Annals and Magazine of Natural History,* 2d ser., 16: 184–196.

———. 1856a. "On the Habits of the Orang-utan of Borneo." *Annals and Magazine of Natural History,* 2d ser., 18: 26–32.

———. 1856b. "On the Orang-utan or Mias of Borneo." *Annals and Magazine of Natural History,* 2d ser., 17: 471–476.
———. 1856c. "Some Account of an Infant 'Orang-utan.'" *Annals and Magazine of Natural History,* 2d ser., 17: 386–390.
———. 1857a. "On the Great Bird of Paradise, *Paradisea apoda*, Linn.; *'Burong mati' (Dead Bird)* of the Malays; *'Fanéhan'* of the Natives of Aru." *Annals and Magazine of Natural History,* 2d ser. 20: 411–416.
———. 1857b. "On the Natural History of the Aru Islands." *Annals and Magazine of Natural History,* 2d ser., 20 (suppl.): 473–485.
———. 1857–1858. Letter [to Samuel Stevens concerning collecting, dated March 10 and May 15, 1857, Dobbo, Aru Islands]. *Proceedings of the Entomological Society of London,* 91–93.
———. 1858a. Letter [concerning collecting, dated Dec. 20, 1857, Amboyna]. *Zoologist* 16: 6120–6124.
———. 1858b. "On the Tendency of Varieties to Depart Indefinitely from the Original Type." *Journal of the Proceedings of the Linnean Society (Zoology)* 3:53–62.
———. 1858–1859. Letter [concerning collecting, dated Jan. 28, 1859, Batchian, Moluccas; extracts]. *Proceedings of the Entomological Society of London,* 70.
———. 1859. Letter [concerning collecting, dated Oct. 29, 1858, Batchian, Moluccas; extracts]. *Proceedings of the Zoological Society of London* 27: 129.
———. 1860a. "Notes on *Semioptera wallacii*" [extract from a letter dated Sept. 30, 1859, Amboyna]. *Proceedings of the Zoological Society of London* 28: 61.
———. 1860b. "On the Zoological Geography of the Malay Archipelago." *Journal of the Proceedings of the Linnean Society (Zoology)* 4: 172–184.
———. 1861. *Ibis* 3: 310–311.
———. 1862. "Narrative of Search after Birds of Paradise." *Proceedings of the Zoological Society of London*, 1862: 153–161.
———. 1864. "On the Physical Geography of the Malay Archipelago." *Journal of the Royal Geographical Society* 33: 217–234.
———. [1869a] 1962. *The Malay Archipelago: The Land of the Orang-utan and the Bird of Paradise. A Narrative of Travel, with Studies of Man and Nature.* London: Macmillan. Reprint. New York: Dover Publications.
———. 1869b. "Museums for the People." *Macmillan's Magazine* 19: 244–250.
———. 1870a. *Contributions to the Theory of Natural Selection: A Series of Essays.* London: Macmillan.
———. 1870b. "Man and Natural Selection." *Nature* 3: 8–9.
———. 1876. *The Geographical Distribution of Animals; with a Study of the Relations of Living and Extinct Faunas as Elucidating the Past Changes of the Earth's Surface.* London: Macmillan.
———. 1880. *Island Life; or, The Phenomena and Causes of Insular Faunas and Floras, Including a Revision and Attempted Solution of the Problem of Geological Climates.* London: Macmillan.
———. 1889. *Darwinism: An Exposition of the Theory of Natural Selection, with Some of Its Applications.* 2d ed. London: Macmillan.
———. 1895. *A Narrative of Travels on the Amazon and Rio Negro, with an Account of the Native Tribes, and Observations of the Climate, Geology, and Natural History of the Amazon Valley.* 5th ed. London: Ward, Lock & Bowden.

———. 1898. *The Wonderful Century: Its Successes and Its Failures.* London: Swan Sonnenschein; New York: Dodd, Mead.

———. 1899. "White Men in the Tropics." *Independent* (New York) 51: 667–670.

———. 1905. *My Life: A Record of Events and Opinions.* 2 vols. London: Chapman & Hall.

———. 1908. Address [acceptance speech on receiving the Darwin-Wallace Medal]. In *The Darwin-Wallace Celebration Held on Thursday, 1st July, 1908, by the Linnean Society of London,* 5–11. London: Burlington House.

———. 1916. *Alfred Russel Wallace: Letters and Reminiscences.* Edited by James Marchant. London: Cassell.

———. 1991. *Alfred Russel Wallace: An Anthology of His Shorter Writings.* Edited by Charles H. Smith. Oxford: Oxford University Press.

Wells, Michael, S. Guggenheim, A. Khan, W. Wardojo, and P. Jepson. In press. *Investing in Biodiversity: A Review of Indonesia's Integrated Conservation and Development Projects.* Washington, D.C.: World Bank.

Whitaker, Katie. 1996. "The Culture of Curiosity." In *Cultures of Natural History,* edited by N. Jardine, J. A. Secord, and E. C. Spary, 75–90. New York: Carmbridge University Press.

Whitmore, T. C., ed. 1981. *Wallace's Line and Plate Tectonics.* Oxford: Clarendon Press.

Whitten, Anthony J., Sengli Damanik, Jazanul Anwar, and Nazaruddin Hisyam. 1987. *The Ecology of Sumatra.* 2d ed. Yogyakarta: Gadjah Mada University Press.

Whitten, Anthony, Muslimin Mustafa, and Gregory S. Henderson. 1987. *The Ecology of Sulawesi.* Yogyakarta: Gadjah Mada University Press.

Whitten, Tony, Roehayat Emon Soeriaatmadja, and Suraya A. Afiff. 1996. *The Ecology of Java and Bali.* Hong Kong: Periplus.

Williams-Ellis, Amabel. 1966. *Darwin's Moon: A Biography of Alfred Russel Wallace.* London: Blackie.

Yerkes, Robert M., and Ada W. Yerkes. 1929. *The Great Apes: A Study of Anthropoid Life.* New Haven: Yale University Press.

INDEX

Design:	Steve Renick
Composition:	Integrated Composition Systems
Cartography:	Reineck & Reineck, San Francisco
Text:	12/15 Adobe Garamond
Display:	Centaur
Separations and prepress:	The Arthur Morgan Company
Printing and binding:	C&C Offset Printing Company

The mission of The Nature Conservancy is to preserve the plants, animals, and natural communities that represent the diversity of life on earth by protecting the lands and waters they need to survive.

Internationally, the Conservancy works through partnerships with governments, businesses, organizations, and communities that share a concern about the escalating rate of extinction worldwide. The Conservancy helps these groups build the capacity to conserve their local biodiversity and natural systems.

Since 1992, the Conservancy's Indonesia Program has focused on developing terrestrial and marine models of protected area management, primarily at Lore Lindu and Komodo National Parks. Proceeds from the sale of this book will directly benefit the long-term protection and management of Indonesia's magnificent national parks.